Robots, Automation and the Innovation Economy

Cascades of new technologies and innovations are entering our lives so fast that it is difficult for us to adapt to one innovation before the next becomes embedded into our everyday lives. What happens when the changes brought by technology are so profound that they affect all aspects of our lives? This book explores the potential impact of artificial intelligence (AI) and intelligent robots on individuals, organizations and society, specifically examining the impact on jobs and workplaces in the future. It provides an understanding of how we can adapt to changes that appear like flocks of black swans.

Five key areas are unpacked in the book: automation, AI, (the significance of AI technology), innovation, competence transformation and the fact that the pace of change is so rapid that it outstrips our ability to adapt to consecutive changes. The main objective is to show how AI will change society and how we as individuals and society must adapt in order to survive what the author terms 'robot shock', together with its consequences and after-effects. It offers a greater understanding of resistance to change and how we need to adopt strategies for adapting to major changes. Each of the book's six chapters also contains policy inputs, framed as propositions, that are intended specifically for decision-makers. The book concludes by offering possible strategies for overcoming the negative effects of 'robot shock'.

The book intends to send a message to leaders of institutions, decision-makers and anyone attempting to understand and explain how we – as a social system – can succeed in tackling the many major challenges and crises faced by humanity.

Jon-Arild Johannessen is a Professor at Kristiania University College, Oslo, Norway.

Routledge Studies in the Economics of Innovation

The *Routledge Studies in the Economics of Innovation* series is our home for comprehensive yet accessible texts on the current thinking in the field. These cutting-edge, upper-level scholarly studies and edited collections bring together robust theories from a wide range of individual disciplines and provide in-depth studies of existing and emerging approaches to innovation and the implications of such for the global economy.

Deglobalization
China–US Rivalry in the Innovation Economy
Jon-Arild Johannessen

Innovation, Automation and a Sustainable Economy
Tackling the Inequality, Climate and Biodiversity Crises
Jon-Arild Johannessen

Decentralized Autonomous Organizations
Innovation and Vulnerability in the Digital Economy
Edited by Sven Van Kerckhoven and Usman W. Chohan

Economics of the Pharmaceutical and Medical Device Industry
Supply Chain, Trade and Innovation
Ramesh Bhardwaj

Innovation and Economic Development
Raja M. Almarzoqi and F. John Mathis

FinTech, Financial Inclusion and Sustainable Development
Disruption, Innovation and Growth
David Mhlanga

Robots, Automation and the Innovation Economy
Jon-Arild Johannessen

For more information about this series, please visit: www.routledge.com/Routledge-Studies-in-the-Economics-of-Innovation/book-series/ECONINN

Robots, Automation and the Innovation Economy

Jon-Arild Johannessen

LONDON AND NEW YORK

First published 2025
by Routledge
4 Park Square, Milton Park, Abingdon, Oxon OX14 4RN

and by Routledge
605 Third Avenue, New York, NY 10158

Routledge is an imprint of the Taylor & Francis Group, an informa business

British Library Cataloguing-in-Publication Data
A catalogue record for this book is available from the British Library

ISBN: 978-1-032-93423-5 (hbk)
ISBN: 978-1-032-93716-8 (pbk)
ISBN: 978-1-003-56732-5 (ebk)

DOI: 10.4324/9781003567325

Typeset in Times New Roman
by Deanta Global Publishing Services, Chennai, India

Contents

Prologue

In this book, we explore the potential impact of artificial intelligence and intelligent robots on individuals, organizations and society.

The underlying problem we address throughout this book is as follows: How will artificial intelligence and intelligent robots affect jobs and workplaces in the future?

The book's starting points are the 2015 open letter[1] about the dangers of artificial intelligence, whose signatories included Stephen Hawking, and the 1970 book *Future Shock* by Alvin Toffler.

Hawking and the other signatories warned that it was vital not to allow the potentially incalculable benefits of artificial intelligence to blind us to its pitfalls. In contrast, Toffler sees the potential dangers of technology, but is generally optimistic about the future impact of technological progress. Toffler's main focus is the accelerating pace of change, while Hawking is more concerned with the inherent dangers of technological change processes. A possible objection to Toffler's stance is that his book predates the development of the technologies necessary for artificial intelligence. However, the world was also going through a period of rapid technological change when Toffler was writing. Toffler's book influenced how millions of people thought about the future. It became a source of inspiration for decision-makers and leaders worldwide.

Many jobs are likely to be automated by artificial intelligence. Potentially, such automation will give us much more freedom over how we spend our time. There is, however, an important question: Will we be able to handle such freedom? Another key question is whether we will be able to tackle the kinds of coordination problems that may accompany this new technology. Could the combination of greater freedom and the challenges posed by coordinating technology cause a reduction in demand for human labour and lead to academics and other people with intellectual competence becoming the new working class?

We reflect on these questions in the various chapters in this book. Each of the book's six chapters also contains policy inputs, framed as propositions, that are intended specifically for decision-makers. The final chapter is concerned with possible strategies for overcoming the negative effects of 'robot shock'. The chapter adopts a practical approach and is oriented towards policy- and decision-makers.

There is a very large difference between humans and robots. The greatest difference, which really makes a difference, can be illustrated in the following anecdote:

We can hit our thumb with a hammer, and we therefore also know how others who do the same will feel. However, intelligent robots cannot do this. The emotional and social level is an important distinction between robots and humans, at least at the current level of technological development.

Core message:
The effects of automation resulting from the deployment of artificial intelligence and intelligent robots are not visible in the unemployment statistics, but in the wage statistics.

Statements of aims

Main themes

Like *Future Shock,* 'robot shock' is intended to send a message to leaders of institutions, decision-makers and everyone attempting to understand and explain how we – as a social system – can succeed in tackling the many major challenges and crises faced by humanity. While 'robot shock' is not 100 per cent optimistic in its view of technology, it is not a dystopian read that adopts a purely crisis mentality in assessing the future. Our general approach is to attempt to balance the two perspectives.

Goals

Our goal is to explore the social mechanisms that are the drivers of robot shock, as well as the cascades of aftershocks that will affect social systems on various (individual, organizational and societal) levels. We attempt to say something meaningful about the short- and long-term effects of these shock waves at these three levels.

How this book differs from other books

Our hope is that this book – just like Toffler's – will change some mindsets. 'Robot shock', like *Future Shock,* employs methods that are relatively unusual in traditional scholarship. We use insights from thought experiments, genealogy, scenario thinking and conceptual generalization. At the same time, we attempt to demonstrate some trends by using small-scale case studies to describe positions or possible changes. We make use of historical insights, artistic approaches, aphorisms, metaphors and literature reviews to emphasize our points. In this way, we emulate Toffler by integrating various methods from different areas to emphasize certain perspectives and points. Readers who are familiar with *Future Shock* will see that our book borrows many of its points and structural elements from Toffler's book. This borrowing was a conscious choice, because few books have influenced our thinking about the future more than Toffler's, plus it is our understanding that we can always learn from the best, even though more than 50 years have passed since Toffler wrote his book.

Objective

One of our objectives with this book is to show how artificial intelligence will change our society and how we as individuals and society must adapt ourselves to survive robot shock, plus its consequences and after-effects.

A second objective is to provide an understanding of how we can adapt to changes that appear like flocks of black swans. It is the degree of innovation and the speed with which these innovations emerge that has the greatest effect on people and society. When we are unable to adapt to an innovation before the next one is launched on the market, this affects how we relate to the changes that the innovations create. How we manage to adapt to 'robot shock' is an essential objective of this book.

A third objective, which is related to the second objective, is to provide a greater understanding of resistance to change and how we need to adopt strategies for adapting to major changes.

The innovative idea in this book

Shock waves are inevitable after an explosion. Once intelligent robots become part of everyday life, one of the aftershocks will involve the emergence of new jobs that will benefit society. Such jobs will require new kinds of competence. In this way, employment will be exposed to creative destruction. When the rate of change outstrips our ability to adapt, people, organizations and society will be affected by these aftershocks. One important point is that if whole systems are exposed to aftershocks, then social sub-systems will also be affected. Another point is that explosions, which is one way in which we can understand artificial intelligence and intelligent robots, will generate shock waves that will have consequences for economic, social, political and relational systems. In turn, these shock waves will generate cascades of aftershocks that will change and influence systems long after the dust of the actual explosion has settled. We explore some of these consequences in this book. The consequences we explore are: automation, artificial intelligence, innovation, competence disruption and the fact that the pace of change is so rapid that it outstrips our ability to adapt.

Methods

We employ both established and innovative methods to deal with the problem of 'the robot shock'. The established methods involve literature reviews, scenario thinking and historical methods. In our context, the historical methods are intended to provide insight into some signals being sent by history to those responsible for making decisions about the present that will influence the future. In this way, we can say that history is both sending signals about events in the past and is also creating expectations about what will happen in the future. The expectations that historical signals convey to decision-makers contribute to the creation of a mechanism based on expectations, which in turn helps to steer current events.

The innovative methods in this book are related to conceptual generalization. In this process, we apply the findings of other researchers and draw associations from these findings to create analytical, conceptual and empirical models, as well as thought experiments, in order to get to grips with the problems and issues under investigation (Adriaenssen & Johannessen, 2016).

We use conceptual generalization as an intellectual tool to tackle the general research problem and the research questions that we examine in each chapter. The tools we apply in conceptual generalization comprise various conceptual models that are designed to provide a coherent general understanding of the structure of each chapter.

We also develop typologies in order to be able to create hypotheses. These hypotheses are then developed as a system of hypotheses, i.e., they are connected in a consistent way. In this way, we are also able to develop a mini theory for the relevant research question. We use Mario Bunge's definition of 'theory', which states: 'A theory is a system of propositions' (Bunge, 1974: v).

We also use Boudon-Coleman diagrams to highlight processes at different levels (micro, meso and macro) for the problem and research questions being investigated. In addition, we use loop diagrams to reveal systemic connections. These intellectual tools are used to come to grips with the various aspects of the problem and questions being investigated.

Note

1 https://en.wikipedia.org/wiki/Open_Letter_on_Artificial_Intelligence

Figures

1 Robot shock

Core idea in this chapter

- The effects of automation resulting from the deployment of artificial intelligence will not be visible in the unemployment statistics but in the wage statistics.

Key points in this chapter

- New technology and cascades of innovations are causing old institutions to collapse and new institutions are arising on the ruins of old ones.
- The Fourth Industrial Revolution, with its innovation economy and cascades of innovations, is generating societal stress and angst.
- Globalization, which has done much to promote industrialization and increase automation rate in China, has also brought about extreme economic inequality.
- The middle class will bear the brunt of the trend towards feudal capitalism.
- If the pace of change outstrips our ability to adapt, we will experience forms of erratic behaviour at individual, organizational and societal levels.
- The possibility of losing something one has already achieved generates a level of resistance that exceeds the energy and resources one would devote to achieving something new.
- In this book, the strategies we have developed for tackling robot shock are designed for local and global applications while at the same time are adapted for application at an individual (micro), organizational (meso) and national (macro) levels.

Introduction

The problem we investigate in this book relates to the development of the workplace of the future (Johannessen, 2020). The underlying concern we address throughout this book is as follows: How will artificial intelligence and intelligent robots affect jobs and workplaces in the future?

The technological drivers we are investigating are new technology, artificial intelligence and intelligent robots. We conduct our investigation by exploring five perspectives on artificial intelligence concerning social systems, as well as a sixth

DOI: 10.4324/9781003567325-1

perspective where we attempt to investigate strategies for survival. In this connection, Stephen Hawking had some very gloomy predictions about the impact of technology on humanity's fate.[1]

The five areas we explore are: Automation, artificial intelligence (the significance of AI technology), innovation, competence transformation and the fact that the pace of change is so rapid that it outstrips our ability to adapt to consecutive changes.

In this book, we explore what happens when the changes wrought by technology are so profound that they impact all aspects of our lives. Cascades of new technologies and innovations are entering our lives so fast that it is difficult for us to adapt to one innovation before the next becomes embedded into our everyday lives.

The best way to change others is to change our reactions to their behaviour (Watzlawick, 1980). But how can we change ourselves when not people but intelligent robots affect every aspect of our everyday lives? This book is about how we should relate to events, technology and social mechanisms that will affect our future in a revolutionary way (Schwab, 2016). Although technology is the driving force behind change, in this book we are most concerned with the human aspect of these changes. One could say that it is 'the human aspect of artificial intelligence' that we are concerned with, not the technology itself (Johannessen, 2022b, 2022d). It is the steps each of us needs to take to future-proof ourselves and our families that we focus on here; that is to say, this book is about how everyday phenomena affect people. Those with a wait-and-see attitude towards the new technology will be 'left behind'. The reason is straightforward: The changes are happening faster than we can imagine and they come in waves. New technology emerges before we become familiar with the latest changes (Johannessen, 2024, 2024a). These rapid changes affect every one of us. On top of this, the new technological innovations have already become part of the conflict between China and the US (Allison, 2018). In other words, this book deals with themes interesting to all of us.

The new technological changes that will emerge in the future can figuratively compare to a roaring flow of lava from an erupting volcano, which will completely change our lives. For some, these changes will lead to greater freedom and a comfortable life. However, for the vast majority, the new technology will dramatically change the workplace. The cascades of innovations will roll out, resulting in increased automation processes, which will become part of our daily lives. In the Fourth Industrial Revolution, artificial intelligence will be integrated into technological products and devices – everything from children's toys to the tools and appliances that adults use in their work and everyday lives (Schwab, 2016).

People's skills will quickly become obsolete and no longer applicable in the new labour market (Johannessen, 2024, 2024a). This skill disruption will affect us all. However, the technological revolution will develop slowly and can perhaps be more appropriately termed 'technological evolution'. Consequently, people will have the time and opportunity to gain new and additional skills compatible with modern technology. Although most of the changes will take time to permeate social systems, there will nevertheless be cascades of innovations entering the market. These innovations will come with such speed that people will have problems adapting to such a fast-changing environment. They will have to adjust to new changes, even if they haven't yet adapted to the less recent changes. This will affect people both psychologically and socially. Regarding organizations, fast-changing environments will result in systems behaving erratically. These systems are unable to adapt adequately to the innovations, and as a result, a wastage of energy will occur. The people, organizations and societies that survive are the ones that can adapt rapidly to the new technology that will continuously emerge in new forms and within new areas of use.

The new technology and cascades of innovations are causing old institutions to collapse, and new institutions are rising upon the ruins of old ones. We can imagine that a 'Phoenix bird culture' will become part of the latest online culture. For example, new technologies result in an increase in shopping online and an increasing tendency for people to make new friends virtually. In other words, the online culture will develop further. With this development, close friends and family will take on a new meaning in our lives. By this, we mean that these institutions, with their new norms and values, will be built upon the ruins of the old institutions. An interesting point in this Phoenix bird culture is that many norms and values will exist long after they cease to have any real function (Bunge, 1996, 1998, 1999). In this way, the slow-moving course of history will influence our behaviour in the here and now, even if we do not know why we maintain such behaviour that is no longer appropriate. In this development, subcultures will become part of people's reality to enable them to cope with the drastic changes that occur during and after the 'robot shock'. These subcultures will be able to act as a form of psychological and social safety net for many people. These safety nets will give people some meaning to their lives and help them to interpret the significant changes that will occur.

When our values and norms change and artificial intelligence invades our lives, few people are interested in considering these changes from a long-term historical perspective. Most people will be more interested in understanding the new situation and what they need to do to cope (Seligman, 2011). It is usually only historians, social scientists and other academics who are interested in considering the major historical perspectives. For most people, it is about survival when cascades of innovations impact their lives with catastrophic consequences. In this context, we will develop strategies to help people survive the imminent 'robot shock'.

When significant technological changes impact a society, this can be compared to a fast-moving river after the spring snowmelt. If we study a river's currents, we can observe that the faster the river rushes forward, the more there will be places where some of the water will end up in backwaters, that is, parts of the river not joining the current. Metaphorically and in a social context, these 'backwaters' will not become part of the new developments or progress but will instead develop their subcultures with their norms and values. The 'backwaters' will become part of the consequences of the major changes, where some people will find a way of explaining and understanding what is happening around them. This situation is on par with people who resort to astrological explanations to understand the world around them. They can continue to do this without anyone being able to point out errors in their assessments. In other words, when phenomena, such as social systems, dreams, astrology and so on, are very complex, one interpretation or theory about them is often as valid or non-valid as any other. In this context, we can understand how people will react to the significant changes that will follow from the upcoming technological revolution – when new AI developments will change all our assumptions and premises and what we have learnt about the world. When the complexity is great, all theories can become valid, because one can select a small part and make it the principal component of the whole. When we do this, we have also given the complexity a purpose, and we understand the complexity through the simplification we have chosen. Regarding dreams and dream interpretation, any charlatan can appoint themselves as 'the high priest of knowledge' because any analysis is possible, even though none are 'right', and none can either be confirmed or denied. We bring this up here because the same applies to the increasing complexity of technological revolutions. This situation is where subcultures gain meaning and give many people a purpose in their lives, similar to theories of dream interpretation and astrology.

Artificial intelligence and intelligent robots are drivers of change in social systems. At the same time, processes in social systems change these technologies. In this way, the changes accelerate and reinforce the development of various subcultures. Technological innovations thus become a central part of the changes that create our everyday lives. Technological advances have both psychological and social consequences. These consequences affect our lives in many ways. For instance, which education should we choose, what should educators teach in our schools, how should we develop our cities and what are our expectations, dreams and future needs and wishes? It is primarily artificial intelligence, and all the applications of this technology, that will drive the future change processes in our society (Schwab, 2016; Johannessen, 2020a, 2022). Some experts are worried that artificial intelligence will get out of control and that the rapid pace of its development will cause social harm. This fear has been expressed by key figures, such as Stephen Hawking and several others, who signed an open letter describing the possible dangers associated with the unfettered future development of AI technology.[2] In this context, we can say that Hawking and the others represent a neo-Luddite perspective towards the new technology. The point is that people, organizations and society need to learn how to live with this technology and quickly adapt to the changes the technology will create. On the other hand, it will also be wise to regulate and control technological development as far as possible. We should note that all the industrial revolutions in the past developed in such a way that some people felt they were losing control over their lives.[3] In other words, it will not be any different this time around either. However, the new aspect of this technology is that it will affect everyone and not only the workers on the manufacturing floor who have been replaced by automation technology. The new AI-driven automation technology will also replace many other kinds of jobs. To a greater extent than before, the new technology will affect the jobs of knowledge workers. In other words, the knowledge workers will be subjected to changes in the workplace in much the same way as the industrial workers were in the previous industrial revolutions. In the Fourth Industrial Revolution, AI technology and automation will result in significant challenges for everyone, unlike the previous industrial revolutions which only impacted specific groups of workers.

The Fourth Industrial Revolution, with its innovation economy and cascades of innovations, will generate societal stress and angst. However, the situation will not get any easier if significant players decide that the latest AI technology should be kept under lock and key until it is possible to control it. This idea seems to be the perspective of Hawking and the other AI experts who signed an open letter warning of the possible negative social impacts of AI.[4]

We argue for a 'double strategy' here: Control and adaptation. Those who can recognize the importance of the new technology more quickly than others, develop it and adapt to its consequences will have an advantage over those who spend time reflecting on the technology's negative consequences. Of course, there will be negative consequences from this technology. This has always been the case with all new technologies, from the invention of the wheel to the invention of nuclear technology and now artificial intelligence as well. Social anxiety and psychological stress have always followed in the wake of new technology.[5]

When the rate of change is significant and machines take over people's jobs and create uncertainty about people's futures, many people become disoriented and seek security in subcultures that can give an unambiguous meaning to the purpose of their lives. Responding to this, the new 'high priests' offer simple and easily understandable solutions. When the latest technology, artificial intelligence, becomes part of our everyday lives in such a short time, social anxiety will spread in social systems. Some people will quickly adapt to the changed circumstances and will be able to experience job satisfaction and consequently flourish (Seligman, 2011). Other people will develop 'learned helplessness' (Seligman, 2006) and become critical of the new technology. The 'fear' of modern technology may become so great that social symptoms may develop linked to these new psychological concepts, and emerging help services may sprout to tackle the symptoms. We can imagine that digital neurosis will become one such psychological concept and that various treatments will surface to help alleviate the symptoms of people 'suffering' from digital neurosis. Digital neurosis will be related to the logic of digital systems, constructed around 1 and 0, that is, binary digits. In other words, people suffering from digital neurosis will, metaphorically, only be able to understand a single value of either 1 or 0. So, they will have a black-and-white perspective and only be able to understand a problem in one of two ways, where all shades of grey will not be possible. Digital neurosis will also be linked to the rise of subcultures, because black-and-white thinking will also abound in these subcultures.

'Robot shock' is not something that will occur in the future; it already exists today, albeit to a lesser extent and with less force than what we might expect in the time to come. In other words, robot shock is active in our communities today. Digital neurosis and other psychological conditions resulting from the new technology are not of the future but are already present today. For example, more and more people are thinking in terms of 'black' or 'white'. Digital neurosis is an outcome of the digital revolution, where everything has a value of zero or one (0–1). In such a world, 'either-or' can quickly become how people think. Nuances, complexity and rational thinking in such a world can rapidly come into disrepute.

We may have little competence, either individually or at a system level, about how to deal with changes that happen very quickly and that come in cascades, like a flock of 'black swans'. These black swans come as a surprise and have considerable consequences. The COVID-19 pandemic (2020–2022) is the latest example of a black swan that had global effects. When changes come suddenly and surprisingly, we have not accumulated enough knowledge to adapt to such changes. It is precisely flocks of black swans that we envision will be the consequence of the development of artificial intelligence and intelligent robots and which will be at the core of the 'robot shock'.

How can we prepare for situations which we do not know will occur? How are we supposed to imagine problems we don't know anything about? This situation is a whole new domain of knowledge we know little about. We term this domain of knowledge 'hidden knowledge'. Hidden knowledge has been described by Kirzner (1982) and Johannessen (2022d). Regardless of whether this domain of knowledge is partially illuminated, we have an extremely poor insight into hidden knowledge. You could say that this domain of knowledge deals with 'what we don't know, that we don't know'. We can assume that this is where creativity and innovation spring from. Three other knowledge domains relate to change processes of which we have a relatively adequate understanding. Explicit knowledge is knowledge that can be articulated, codified, stored and accessed. Implicit knowledge is 'what we know we do not know'. It is mainly this area of knowledge that research focuses on. Tacit knowledge is knowledge that is gained through experience, that is difficult to convey to others (Polanyi, 1966, 1969, 2015). Of the four knowledge domains, hidden knowledge will be the most relevant when we need to adapt to flocks of black swans. This knowledge provides insight into creative processes that lead to innovation. The innovation processes of the Fourth Industrial Revolution will create flocks of black swans.

When major changes occur, it seems that most of us react irrationally, without any appropriate strategy to deal with the changes. However, much research has been conducted in this field, i.e., how to cope with changes in the best possible way. Of particular interest is the Prospect Theory of Kahneman and Tversky (1979, 2000). This theory has been described and elaborated on by Adriaenssen and Johannessen (2016a), amongst others.

There often appears to be irrational and blind resistance to change. This particularly applies to changes linked to black swans, and, not least, flocks of black swans. We have indicated areas of knowledge that can give us some understanding of what drives these changes and the resistance to change. The first is the concept of hidden knowledge. The second is Prospect Theory, which tells us something about the irrational forces that resist change.

One of the purposes of this book is to show how artificial intelligence will change our societies. Another purpose is to develop strategies at the individual, organizational and societal levels that can help us survive 'the robot shock', its consequences and its aftermath. In Chapter 6 of the book, we develop strategies for how we can adapt to changes and live with a reality where flocks of black swans invade our lives and our societies.

The extent and rate of innovation diffusion will determine the range of the impact on people, organizations and society. When people are unable to adapt fully to a major innovation before a new one emerges, this affects how they relate to the changes resulting from the innovations. Therefore, how people will adapt to 'the robot shock' is one of the essential purposes of this book.

The strategies for adaptation to the robot shock that we develop in Chapter 6 are, in reality, a theory for adjustment when cascades of innovations flood the market. The theory is a step-by-step method of how people and society can deal with robot shock. This idea is an innovative part of the book's practical theory development.

We employ both established and innovative methods. The established methods involve literature reviews, scenario thinking and historical methods. In our context, the historical methods are intended to provide insight into some signals being sent by history to those responsible for making decisions about the present that will influence the future. In this way, we can say that history is both sending signals about events in the past and is also creating expectations about what will happen in the future. The expectations that historical signals convey to decision-makers contribute to the creation of a mechanism based on expectations, which in turn helps to steer current events.

The more innovative methods in this book are related to conceptual generalization.[6] In this process, we apply the findings of other researchers and draw associations from these findings to create analytical, conceptual and empirical models, as well as thought experiments, in order to get to grips with the problems and issues under investigation (Adriaenssen & Johannessen, 2016).

We use conceptual generalization as an intellectual tool to tackle both the general research problem and the research questions that we examine in each chapter. The tools we apply in conceptual generalization comprise various conceptual models that are designed to provide a coherent general understanding of the structure of each chapter.

We also develop typologies in order to be able to create hypotheses. These hypotheses are then developed as a system of hypotheses, i.e., they are connected in a consistent way. In this way, we are also able to develop a mini theory for the relevant research question. We use Mario Bunge's definition of 'theory', which states: 'A theory is a system of propositions' (Bunge, 1974: v).

We also use Boudon-Coleman diagrams to highlight processes at different levels (micro, meso and macro) for the problem and research questions being investigated. In addition, we use loop diagrams to reveal systemic connections. These intellectual tools are used to get to grips with the various aspects of the problem and questions being investigated.

There are some reservations that are important to clarify regarding this book.

First reservation: This book was written over a number of years and has been rewritten several times. Consequently, some of the facts that are referred to will perhaps be 'out-of-date' by the time the book is published. We have tried to minimize this by using cases and case letters that are more general than specific. An important point about generalizing examples is that they become less concrete. Nevertheless, despite this generalization, some of the examples, cases and case letters may still seem 'out of date' and irrelevant in the new context in which the book is read. We are aware of this and thus recommend that the reader takes this time lag into account. In this context, the time lag relates to two factors. First, a problem is examined at a certain point in time and may be less relevant by the time the reader reviews it in a book dealing with the problem. Second, it takes time from when a book proposal is submitted to a publisher, until the book is available; this is a relatively lengthy process and can take from one to two years. It is these two factors relating to 'time-lag' that we ask the reader to take into consideration when reading the book.

Second reservation: It will always be problematic using the verb 'will', when expressing the future tense to talk about an event in the future. This was a problem that was also addressed by Toffler (1970: 5). In other words, we do not intend to make 'predictions' when writing about 'the robot shock', but will formulate hypotheses about what may happen under given circumstances. We will leave predictions about what will happen in the future to astrologers, fortune tellers and the like. Instead, we will focus on social mechanisms, systemic connections and what-if connections. Scenario thinking, or methods related to forecasting, has such great complexity that we must be careful in saying anything concrete about a future no one knows anything about. On the other hand, we can investigate how social mechanisms can trigger social processes, which in turn can provide insight into possible outcomes of the robot shock and other possible futures. We are also aware that when the complexity is great, this will create a situation where various 'charlatans' enter the scene and present themselves as the new 'high priests' who claim they can predict the future.

In this book, when we examine social mechanisms that drive social processes, this will always be subject to uncertainty and doubt. Therefore, in every context we should use conditional words, such as 'if', 'whether', 'probably' and so on in order to ensure that the reader understands that what happens will depend on various circumstances and assumptions. However, such a writing style, characterized by an overuse of conditional words, would make the writing difficult for readers to follow. Therefore, we believe that a capable reader will understand that when we write in this book that such and such social mechanisms will lead to such and such consequences, such assumptions always imply uncertainty. The point here is that a lack of certainty regarding different possible outcomes should not deter us from investigating the possible outcomes of the introduction of new major technology such as artificial intelligence. In this context, we can refer to Stephen Hawking, who, along with other co-writers, wrote in an open letter that artificial intelligence can pose a considerable danger to humanity, and that the research, development and implementation of the technology should not progress unhindered. It is commendable that such prominent AI experts openly present their views on this new technology. Others, however, may have a more optimistic viewpoint regarding the future social consequences of the development of AI.

Of course, it is best to have the hard facts, so we can discuss issues with confidence. However, even if we don't have hard facts, that doesn't mean we shouldn't be able to examine the possible outcomes of future actions and events. In this context, we can learn something from 'The Father of History', Herodotus,[7] who used systematic investigation of historical events to tell us how we should act in both the present and the future. Where hard facts are not available, we can investigate a phenomenon or problem by using thought experiments, assumptions and theories and try to address the possible future consequences, even if we cannot relate this to hard facts. On the other hand, even if we have the hard facts, different perspectives and interpretations will give different answers to the same question. In this way, 'hard facts' will not be sufficient to tackle the possible consequences of the development of artificial intelligence.

Assumptions and notions do not have to be correct to be useful. They can spur on reflection and further investigation, even if they do not point towards what would later prove to be the exact outcomes. Even when researchers make mistakes, their studies can lead in the right direction. Making mistakes is thus not necessarily injudicious but can, in the long run, provide fruitful answers.

This book is a map that can shape and change the terrain. The book is not, as many researchers and students often view scientific research, about drawing maps that match the terrain. Often when we investigate something about possible futures, we know little or nothing about what it looks like. On the other hand, the maps we draw can help shape the future terrain. In most cases, others who receive new information about how social mechanisms affect possible outcomes will continuously reshape the landscape. Thus, Hawking and the other open letter signatories may turn out to be right in their assumptions about the future of AI. But it may also be the case that because they pointed out the dangers of what they assumed could happen, it then doesn't happen. In this way, the consequences of making what seems to be a reasonable prediction can result in the opposite happening and vice versa. Consequently, hypotheses and theories, even if they are incorrect, can be right in their consequences.

We have structured this introduction in Figure 1.1, which also shows how the rest of this chapter is structured. Figure 1.1 also shows how the entire book is structured.

In this book, we will examine the following general research question: What is the potential impact of 'robot shock' on future employment trends?

In order to address the general research question, we have developed the following research questions:

RQ1 What is the potential impact of automation on future employment trends?
RQ2 What is the potential impact of artificial intelligence on future employment trends?
RQ3 What is the potential impact of innovations on future employment trends?
RQ4 What is the potential impact of transformation in competence on future employment trends?
RQ5 What is the potential impact on future employment trends of the fact that the pace of change may outstrip our ability to adapt?

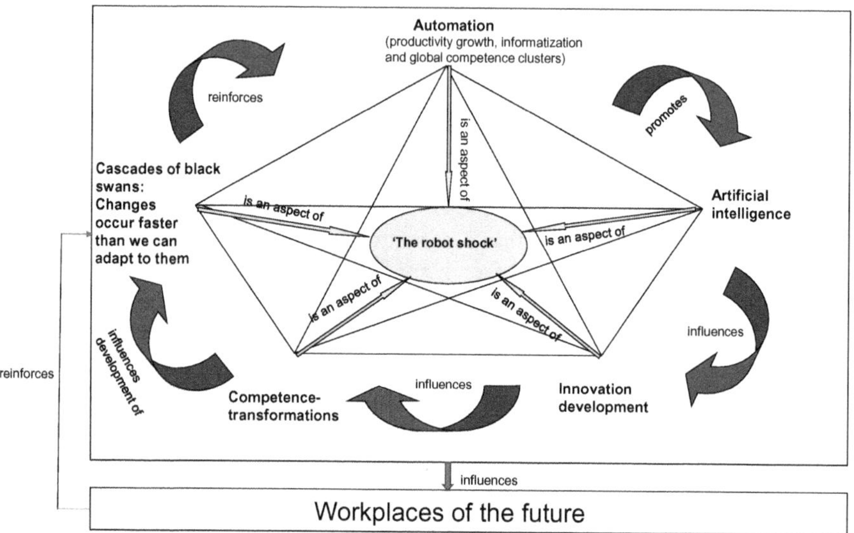

Figure 1.1 Robot shock.

For the rest of this chapter, we will examine each of the six elements of the model. In the remaining chapters, we will examine the elements in more detail, using descriptions, analyses, theoretical reflections, reviews of practical utility and the systemic connections between the elements.

Automation

Automation is nothing new. Perhaps we can go back to the invention of the wheel to understand some of the consequences of automation. The first wheelbarrow made manual labour less tiring. In other words, historically, the toil of manual labour has been a driving force behind automation processes. However, it was during the digital revolution that automation processes began to replace the drudgery of much mental work.

The First Industrial Revolution is often described as a turning point in the development of automation (Skidelsky & Craig, 2020: 1–7). New technology led to a continuous stream of automation processes. In this development, some ended up as losers, and some as winners, in the short and long term. This development may be related to the concept of skills. Old skills became redundant, while new skills increased in demand. Thus, the social consequence of this new development was that people had to learn new skills to adapt to the latest technology.

A central question today, which was also an important question during the First Industrial Revolution, is whether workers are better off due to the automation that follows industrialization. Regarding the First Industrial Revolution, the observable answer is that people benefitted from industrialization in the long term, but that some benefitted more than others (Frey, 2020). We see a similar tendency with today's automation because of digitalization. Productivity and profits have increased, but wages have largely remained at a standstill from the 1970s up until today (Aghion et al., 2021). In addition, we see that many people, who have not been able to acquire new skills, have been 'left behind' in an economic backwater. Such a development is observable in the US, where feelings of alienation (Carney, 2020), despair (Frank, 2004) and powerlessness (Frank, 2020) are spreading through sections of the population. In addition, a growing number of 'nomadic' workers travel around the country in search of casual jobs (Bruder, 2017). Moreover, in the US, Europe and other places in the world, there is a growing group of people who do not have permanent jobs but who do on-call work; that is, they are contract workers who go to work only when they are needed (Standing, 2014, 2014a, 2016, 2019). One of the consequences of this development is a large increase in economic inequality and the emergence of new impoverished groups of people such as the 'working poor' (Shipler, 2005).[8] Another perspective is that the new technology and automation have displaced unskilled workers and reduced the possibility of upward social mobility (Eubanks, 2017). This increasing economic inequality is not only a Western but also a global problem (Milanovic, 2016). There are many indications that technology, productivity development and automation are related to a political ideology that reinforces economic inequalities. This ideology was initially linked to neoliberal ideology (Petras & Veltmeyr, 2011). Today, however, we can perceive the contours of a new ideology gaining ground, which we can term 'neo-feudalism' (Kotkin, 2020) or 'feudal capitalism' (Johannessen, 2023).

One of the consequences of the First Industrial Revolution was that the cottage industries were no longer able to compete, resulting in a loss of jobs and income for the cottage workers (Frey, 2019: 11). In the US, a clear consequence of the new technology, digitization and an increasing application of artificial intelligence is a split in the middle class (Reeves, 2018) between a prosperous segment, who move socially upwards, and an impoverished segment, who move socially downwards and become part of 'the precariat' (Standing, 2014, 2014a) or who remain in the middle-class but are low-paid. Similar developments are occurring in Europe. On the other hand, the middle class in China is growing, both in number and income.[9] Nevertheless, it seems that the middle class, both in China and the West, is becoming structured along 'feudal' lines (Kotkin, 2020) and possibly also becoming part of a new 'feudal' capitalism (Johannessen, 2023).

In the US and Europe, we can observe a development where many people who could potentially have become part of the middle class in the past have nevertheless ended up with low-paid jobs (Frey, 2019: 11; Ainley, 2016). Against this background, we can say that the effects of automation resulting from the deployment of artificial intelligence and intelligent robots are not visible in the unemployment statistics but in the wage statistics.

Until globalization gained strength around the 1980s, when neoliberal economic ideology became a reality, industrial workers in the US and Europe could expect to have an income that was on a par with many middle-class workers. However, this development came to a standstill during the period of globalization and neoliberalism (Slobodian, 2020; Piketty, 2016). The consequences in the US and Europe of this economic policy was lower social mobility – it became harder and harder for the working class to climb up the rungs of the social ladder. It is in this context that we use the concept of a development towards feudal capitalism (Johannessen, 2020, 2020a, 2021a, 2022c, 2023).

The consequences of industrialization and automation can be understood in relation to both short-term and long-term perspectives. From the short-term perspective, creative destruction will lead to many social crises for people who have to change professions and perhaps move to a new unfamiliar location (Aghion et al., 2021). In the long term, there will be more positive consequences for most people, both financially and socially (Frey, 2019, 2020). However, in both time perspectives, there will be winners and losers. For some, temporary bottlenecks will halt their social progress. For others, the bottlenecks will only act as an adaptation mechanism to enable them to develop new skills and thus enjoy the benefits of automation.

In every period of industrialization, during the First (the steam engine), the Second (science, electricity and mass production), and the Third Industrial Revolution (digital technology), it has been the local areas where the consequences have been felt most strongly. This was particularly evident during the First Industrial Revolution, when the spinners and weavers of the cottage industry were hit the hardest (Mantoux, 1961; Gaskell, 1833). Their livelihoods were largely ruined. Consequently, most had to move to urban areas to find work and make a living. A similar industrialization process occurred in the 20th century, when the industrialization of China gained momentum from the 1980s onwards, and when local industrial areas in the US, Britain and other industrialized countries suffered drastic negative social consequences. Industrial jobs were outsourced to China, and workers in the US, the UK and other developed countries often had to find new jobs such as low-paid service jobs. In the US, this led to feelings of powerlessness (Frank, 2020) and alienation (Frank, 2004) spreading through the population. We also saw similar developments in Europe, such as in the UK (McGarvey, 2018). In the US, in addition to many people having to do lower paid jobs, the real wages stagnated at the level of the 1970s (Acemoglu & Autor, 2011; Chetty et al., 2017: 398–406). Moreover, a good education, the social mechanism that traditionally enabled working-class and middle-class people to move socially upward, has become less and less accessible (Ainley, 2014, 2016).

Globalization, which has done much to promote industrialization and increase the rate of automation in China, has also brought about extreme economic inequality (Milanovic, 2016). However, it is not only economic inequality, poorer wage growth and feelings of powerlessness and alienation that have become apparent during the period of globalization, from the 1980s to 2022, when decoupling processes took effect. During this period, crime also increased, and health challenges became more apparent. Moreover, in the West, middle-income people find it harder to enter the housing market. During the same period, the divorce rate has risen sharply. In addition, the suicide rate has increased sharply, and there has been a significant increase in mortality due to alcoholism (Frey, 2019: 11). There is, of course, a difference between causal relationships and arbitrary developments over a given period, so it is difficult to point out specific causal relationships. Moreover, those with lowered living standards are possibly not so interested in causal relationships, but focus more on the fact that they have ended up in a negative financial and social situation, a situation not of their own making.

In the US, the consequences of globalization have led to many people who previously voted for the Democrats switching to the Republicans and voting for populist candidates such as Donald Trump. We have seen similar developments in Europe. Consequently, it does not seem to be the case that lower wages and poorer prospects have led to people veering towards socialist politics but rather the opposite: There is an increased tendency for people to vote for right-wing populists.[10] The disintegration of the middle class in the West has thus also led to less emphasis on the traditional structure of the left–right political spectrum.

However, it should be emphasized that the consequences of automation are difficult to determine causally. On the other hand, as we have indicated above, one can say something about the consequences over a period of time, especially the short-term negative consequences of industrialization and automation. It is possible that such a description is just as important and real for those people who are exposed to the negative consequences of automation.

Artificial intelligence

Artificial intelligence will affect future working life in two ways. First, old skills and old jobs will disappear (Johannessen, 2024a). Second, new skills and new job opportunities will be created by the new technology (Johannessen, 2024). Like previous major technological breakthroughs, people, businesses and society will have to radically adapt in order to cope with the major changes.

From the 1980s onwards, there has been a dramatic increase in robotics and industrial automation (Kochan & Dyer, 2021: 58). This development occurred at the same time as the globalization we have described above. This temporal relationship was more arbitrary than causal. From the 1980s onwards, information and communication technology and the internet became the primary leading technologies. At this time, artificial intelligence was still at a very early stage of development. Automation, as a result of ICT and later the internet and increasing digitization, affected people's skills in many ways. People, businesses and society had to adapt to changes, and they became more aware that adaptation to changes was a salient prerequisite for economic progress. The new technology that became a reality from the 1980s onwards automated workplaces in two ways. First, technology increases the productivity of established work processes. Second, technology paved the way for new work processes to emerge. These changes led to a demand for new skills and competencies, such as in network design and data security.

Industrial production first introduced robots to take over difficult and dangerous jobs that exposed workers to high risks. For instance, in the 1980s, the automotive industry used robots to paint car bodies. Furthermore, some industries employ robots to automate workers' heavy work tasks, which could lead to wear and tear and occupational injuries. For example, warehouse robotics enabled the automation of heavy work tasks and also included packaging and delivery operations. However, at this point in time, artificial intelligence did not play a decisive role in these work operations, although a good deal of machine learning was involved (Ford, 2016: 17).

Artificial intelligence can be defined as the ability of machines to learn and make rational choices. For example, chess computers that have been able to beat the world's best chess players (Kurzweil, 2008).

The initial development of artificial intelligence in the 1950s was related to cybernetics and machine learning technology (Simon, 1965). Key researchers in this field were Turner, Wiener and Shannon, amongst others (Simon, 2019). One of the first computer programs that was designed was able to play checkers against a human opponent. However, after many years of development, inventors designed computer programs that could 'learn'. The creation of the two chess computer programs, AlphaZero[11] and Stockfish[12] proves this. AlphaZero is based on a learning algorithm and programmed with the rules of chess; by playing against itself in simulated games, it can find the best moves. On the other hand, Stockfish is a rule-based chess computer program based on existing programmed knowledge of previous chess games played by the grandmasters. While AlphaZero uses large amounts of data and 'learns' by playing against itself, the Stockfish program uses existing knowledge as a basis. In a game between the two computer programs, AlphaZero beat Stockfish precisely because it has the ability to 'learn'.

Today, AI programs have been developed for language translation, facial recognition, analysis of trends, pattern recognition and analysis of consumer needs. Consumer needs programs often know what a customer wants long before he/she is even aware of it. In addition, AI can do gene-editing experiments (Carey, 2019). AI can also do problem-solving and decision-making, a field of study in which Herbert Simon has conducted much fruitful research (Simon, 1965).

Another development that has followed in the wake of new automation technology is the transition from automating production processes to automating information and communication processes. In other words, automation technology first changed the working conditions of 'blue-collar workers', but is now changing the working conditions of 'white-collar workers'. In addition, AI applications are being increasingly used to automate work processes where knowledge workers were previously dominant. If this development continues, it will soon only be the creative and innovative professions that have not been impacted by AI technology. There has been little focus on artificial intelligence, and how it will affect the development of knowledge and innovation work, but this is nevertheless important. It is important because, historically, we have witnessed how blue-collar workers were first affected by automation processes, followed by white-collar workers. Thus, it seems reasonable to assume that new technology will also impact knowledge work in the future. The rationale is that the orientation of artificial intelligence is towards information, communication and knowledge processes. If this assumption is correct, future workplaces will become more or less fully automated. Such a development could have significant consequences for skills development, people's settlement patterns and the development of urban areas. The reference here to fully automated workplaces is to make a point. In reality, there will always be people who work. In this context, we can refer to the well-known witticism: '*In the fully automated factory, there will be one human and one dog. The dog is there to protect the machines, and the human is there to feed the dog*'.

However, this development does not mean that unemployment will grow or become a problem because there will always be new job opportunities in service and completely new industries, which every new industrial revolution has produced. On the other hand, automation can put pressure on wages. If this occurs as prices rise, social conflicts may develop in many countries.

Artificial intelligence has already led to many areas of knowledge gaining access to technology that facilitates the streamlining of work. It is not only chess programs that have utilized AI in learning processes. The development of AI technology has also impacted the translation industry. Today, many translation tools, services, agencies and companies use AI applications. One such service is Google Translate, a multilingual neural machine translation service. Another language service provider is Lionbridge, which offers translation, interpreting and other linguistic services. However, there are many competitors in this market, such as Lingotek, Hogarth Worldwide, Appen Crowdsourcing and MotionPoint (as of 2022). This niche is a growing industry, and the people who will be in demand are those experts in programming and algorithm development. In other words, the skills most probably needed in the future are those compatible with the new technology.

It is not the case that people's skills will become unimportant in the new era where artificial intelligence will be the leading technology as we head towards the Fourth Industrial Revolution. It is how this expertise will be in demand and at what price that is of interest. Part-time work, i.e., 'jobs on demand' and a reduction in permanent positions, will characterize the future workplace (Johannessen, 2024, 2024a). We have already seen the flourishing of one-man and smaller businesses that sell their expertise to larger companies. Many argue that this increases the freedom of the individual, because he or she can be his or her boss and does not have to have someone above them in the hierarchy. However, when workers become heavily dependent on one or a few large companies, a discussion ensues about whether the worker is more 'free' or becomes 'unfree' in a new way. In addition, it is easier for companies to determine the price and conditions of work when they can play several subcontractors against each other. This results in increased competition for service fees and lower incomes for sub-contractors in such a gig economy (Kessler, 2018; Woodcock & Graham, 2019).

New technology increases the automation of work processes in two ways. First, during a transition period, it is used to improve existing work processes. This leads to the old ways of working being maintained but by using the new technology. Second, the way of organizing the work will eventually be replaced, so that the application of the new technology will become optimal (Frey, 2020). Probably, we will observe a similar development regarding the increased use of artificial intelligence. If this is a correct assumption, we can expect a sort of 'reverse ketchup effect' in the job market when work processes change qualitatively, and artificial intelligence eventually replaces human competence. As a point of illustration, we can consider the importance of AI in road traffic. Currently, AI technology serves as an aid for drivers. However, when there is complete technological development and the legal aspects of the consequences of the technology are ironed out, we will probably see such a 'reverse ketchup effect' for workers in the transport sector. In practice, the drivers will fall out of the labour market, and artificial intelligence will take over their jobs. This situation will, of course, not apply to all the jobs in the transport sector but to many jobs, so it is very probable that artificial intelligence will impact the development of social systems in a specific way regarding this sector.

Innovation

Innovation has a Janus face.[13] On the one hand, innovation promotes the development of various aspects of our workplace. Work becomes more simple and safer, and the injuries from heavy manual work disappear. On the other side of this Janus face are the negative aspects of innovation. Some innovations destroy the old working methods and processes and also many professions and trades. For example, during the First Industrial Revolution, innovations, such as the machines used in the textile industry, enabled large-scale manufacturing in urban areas, but it also resulted in the destruction of the domestic industry of spinners and weavers. These domestic workers often had to move to the towns and find jobs in the new factories. The working conditions in the factories and the living conditions in the towns were often very poor (Gaskell, 1883; Mantoux, 1961). In other words, innovation is not necessarily beneficial for everyone, especially for these textile workers.

Another way of viewing innovation is from both short-term and long-term perspectives. In the short term, we can argue that the negative aspects of innovation are a reality for many people. From a more long-term viewpoint, however, it can be said that the same social groups, not the same individuals, prosper much better financially in their new work functions (Allen, 2009; Mantoux, 1961). Against this background, it seems reasonable to assume that similar social mechanisms will also apply in the Fourth Industrial Revolution (Schwab, 2016, 2018).

Investigating the consequences of innovation is like writing a story under continual development because we have to consider the Janus effect, that is, the short-term and long-term perspectives of the consequences; otherwise, the answers we will arrive at will be unreliable and ambiguous.

When I have previously written about innovation leading to economic crises, this is both right and wrong at the same time (Johannessen, 2018). From the short-term perspective, this is correct because many jobs and work functions will disappear, and many people will become unemployed. From the long-term perspective, this is incorrect because many jobs that will have disappeared were dangerous and harmful. In addition, the new work functions create increased productivity, which can provide greater financial well-being for many people. Like other stories and tales, the story of innovation includes 'multi-headed trolls', which can be difficult to explain rationally. In the workplace, to gain insight into the consequences of innovation, it can be advantageous to refer to the Janus effect and the different time perspectives if we are to say something about how the workplaces of the future will be affected by innovations. The context of innovation we are talking about relates to artificial intelligence and intelligent robots. In other words, in this book, we will examine how the consequences of 'the robot shock' will affect future workplaces and not specifically examine innovation or artificial intelligence as a technological phenomenon.

Innovation can be understood as a conceptual tool for understanding different types of changes. These changes can be related to institutional, market, service and technological innovations, to name a few. It is mainly technological innovations that we are concerned with in this book, in particular the new applications of AI and robotics.

In the Fourth Industrial Revolution, it will be expertise in this core technology that will be decisive for taking the lead in the global competition (Johannessen, 2020a, 2020b, 2021b, 2022). It is also this core technology that will become decisive in the competition and the build-up of tensions between the US and China (Johannessen, 2021, 2022a). In this perspective, innovation is not something that is neutral, something that everyone is interested in and which serves the interests of everyone. Innovations in general, and technological innovations in particular, will impact people's everyday lives, destroying some jobs and people's competencies, but at the same time, creating new jobs and a demand for new skills (Johannessen, 2024, 2024a).

The development of artificial intelligence and intelligent robots is creating increased tensions between the US and China (Allison, 2018). Consequently, this can change entire supply chains in the global economy. Moreover, the technological innovations we can see emerging, with artificial intelligence as the core technology, may change the type of globalization we have witnessed from the 1980s until approximately 2023. If such a development occurs (and there are many indications that this is the case), then the Janus effect of innovation will indeed become a reality (Johannessen, 2022a).

Innovations first operate as a creative process, which, for many people, functions as a conceptual tool. Then this technological development becomes a 'game', a recreation, a pastime, for selected people. At this stage, this 'game' does not seem to have consequences for most people or their working lives.

After some time, these technological innovations become more than a game for the few. For instance, the first drones were experimental models; only recently have they been used for practical purposes in various industrial sectors. When an innovation has reached this developmental level, people, organizations and society become dependent on the technology, which we can observe in many industries that utilize drone technology. This dependence may progress into a 'master-slave relationship', where we cannot live without this technology. It controls and takes over our thinking about possible scenarios for the future. In this way, we can compare technological innovations to the development of drug addiction. First, there is an experience of pleasure and increased freedom. Eventually, this ends with the addiction, taking complete control of our lives.

Throughout human history, innovation has been important for work, organization and economic prosperity. This fact is nothing new in our time. What is new, however, is that innovation has begun to rate highly on the geopolitical agenda and has become a significant issue in geo-economic matters. There is a global race to be at the forefront of the development of technological innovations, especially artificial intelligence and intelligent robots. This scenario is evident in several aspects of social systems. The economic aspect is important, but there are several other important reasons. The nation or nations that manage to come to the fore economically with the help of this technology will also be able to be at the fore in other ways. AI and robotics offer a new avenue to achieve or maintain military superiority. Of course, both the US and China are aware of this fact. This can help explain why the tensions between these two superpowers are increasing (Allison, 2018).

The picture that emerges in an overall historical outline from the 1980s up until today is China's economic progress. The second trend is the US's relative, not real, decline. During the last 40 years, China has been transformed from being a backward developing country, with internal conflicts, power struggles and failed economic strategies, to become the world's second largest economy. During the same period, the US involved itself in wars, such as the two wars waged against Iraq; military coups abroad; numerous military and political interventions in Asia, Europe, the Middle East and Africa; and most recently, the 20-year war with Afghanistan and the subsequent withdrawal of American forces. At the same time, the US has begun to focus more on domestic policies, such as racial issues, gun control, immigration and so on. In addition, feelings of powerlessness and despair have spread amongst sections of the population (Frank, 2020; Carney, 2020), leading to disenchantment with the political system and polarization of politics.

Before the Second World War, we can say that the British Empire still had the global lead in terms of GDP and military dominance. However, this situation changed dramatically after the war because the British economy had been devastated, and its empire was quickly crumbling. The British decline as a global power also affected its technology leadership. After the Second World War, a broken Britain passed the baton of global leadership to an increasingly powerful US. In other words, the US now takes the global lead economically, technologically and militarily. Admittedly, the successful Russian launch of Sputnik into space in 1957 came as a 'shock', which seemed to threaten American scientific superiority. This threat, though, was more of a passing phenomenon than a real threat. The internal disintegration within the Soviet Union, which resulted in its dissolution (1988–1991), also meant that it no longer posed a real threat to the US economically, technologically and militarily. After Russia invaded Ukraine in 2022, it seems apparent that Moscow has become Beijing's 'junior partner'. Paradoxically, it might be said that the Russian invasion of Ukraine has strengthened China's global role in relation to the US.

Much has been written about the US's structural economic decline (Atkinson & Ezell, 2014: 85–128). Much has also been written about a supposed inevitable future armed conflict between China and the US (Allison, 2018). In this context, however, one should be aware of the fact that the US is still the world's largest economy and by far the world's strongest military power. A future military conflict between the US and China, which Allison suggests may be a possible future scenario, will not benefit China's economy. Although China has had a rapidly rising economy since the 1980s, it is a fragile system that needs to continuously 'deliver' regarding the expectations of a growing middle class. If China enters into a military conflict, economic stagnation may result, and the growing middle class may become discontent with the Communist Party. The people may cease to support the Party, thereby threatening its durability. However, this is not to suggest that a new political situation will occur comparable to the Tiananmen Square Incident,[14] when troops accompanied by tanks fired at demonstrators. Against such a backdrop, the Chinese leaders will likely do everything possible to avoid global military conflicts that could damage their economic stability and progress. If such an analysis is correct, then in the future, we may see increasing tensions between China and the US. This situation, though, will not trigger a direct military conflict. Underlying these tensions will be the development of technological innovations, where AI and robotics will play a decisive role. Regarding this technological competition between the major powers concerning future workplaces, STEM skills will be in high demand. In the future, new STEM skills will most probably lead to competence transformation throughout the whole of working life.[15]

Transformation in competence

In order to understand the social and economic transformation that is taking place, it is just as important to examine the wage statistics as it is the unemployment statistics. First, real wages in the US and several European countries have stagnated since the 1970s. The reason can be found in globalization trends, where China has supplied the West with cheap consumer goods, because the production costs were lower than in the West. Outsourcing of work to China has harmed US and European employment (Petras & Veltmeyr, 2011). Today, the supply and value chains have changed. Some consumer goods previously produced in China are now manufactured in other low-cost countries such as Bangladesh. In addition, some of the production has been re-shored to the US and Europe. However, the production cost in the West is still higher than in China, resulting in a price rise for some consumer goods. To meet the global competition, the West pushed down wages. In addition, the high-tech competition between the US and China is intensifying, as is the competition between Europe and China. The Chinese high-tech workers have higher wages than the Chinese workers who produce cheap consumer goods for the West. However, the salaries of Chinese high-tech workers are still significantly lower than those of high-tech workers in the West. We can therefore observe a similar competition for high-tech products from China that we saw regarding cheap consumer goods from China from the 1980s to around 2020. The Chinese high-tech products are putting pressure on the competition in the West, and the wages of high-tech workers in the West are under pressure. China's STEM workforce thus poses a challenge for the STEM workforces in the US and Europe, resulting in the wages of Western knowledge workers coming under pressure. This situation will probably continue well into the future if there are no new developments. In this context, one such new undertaking was the US banning the sale of high-tech products made by Chinese companies due to a risk to national security or so the argument went. More specifically, Huawei was restricted from Western markets because of 'national security', thus eliminating a strong Chinese competitor from Western markets.

We have included this reflection because, even if new STEM competencies in China and the West lead to competence disruption, this does not necessarily mean those with in-demand competencies will secure a high salary in a globally competitive market. However, it is also possible that Western knowledge workers can manage to maintain a top salary level if leaders introduce various protective measures. We have already mentioned one such measure, the national security argument. In other words, this may have a positive effect on the salary development of knowledge workers with STEM competencies in Western countries. In addition to the current protectionist measures, it is probable more protectionist and nationalist policies will be implemented in the future. This will result in tougher tariff walls around the West, reducing the competition from the relatively cheaper Chinese high-tech products and services. For a period of time, such protectionist measures will probably be positive for the salary development of STEM workers in the US and Europe.

In addition to the competition for wages between the STEM workforces in China and the West, we can observe a development where graduates with Bachelor's and Master's degrees find it hard to obtain full-time, permanent positions. Moreover, a new class of low-paid workers with insecure jobs is emerging that Standing (2014, 2014a) calls 'the precariat'. These workers, who are part of an emerging GIG economy (Kessler, 2018; Woodcock & Graham, 2019), can't secure jobs at a decent wage. These are underemployed people, although they are well qualified to perform their work. In the US and partly in Europe, we can see that many of these workers are becoming alienated and frustrated and experiencing feelings of powerlessness; this social group often has right-wing populist sympathies.

In the future, competence transformation may be one of the significant social disasters of working life in recent history. Even if students follow their dreams and invest time and money in completing short and extensive university degrees, they may not get the jobs they expected. The jobs graduates end up with are often poorly paid and very insecure (Ainley, 2014, 2016).

During the post-war period, people perceived education as the key to social mobility, enabling young working- and middle-class people to climb the economic and social ladder. If this way up in the social system becomes blocked (and closed in some regions and sectors), then it seems reasonable to say that a social disaster could be just around the corner. Why should young people complete university studies if they only end up having large student loans and poor prospects? Throughout the post-war period, from 1945 until sometime in the 2000s, most parents have hoped that their children would be better off than they were. However, this hope seems shattered, contributing to a future social catastrophe noted by many critics (Ainley, 2014, 2016).

Competence transformation will probably secure the livelihoods of those with extensive high-tech education compatible with the new technology. On the other hand, there is a lot of evidence that those without such STEM competence will only be able to find poorly paid insecure jobs. In addition, it looks like part-time work is becoming the new norm (Kessler, 2018; Autor, 2015).

If we look at competence transformation from a macro perspective, it is related to price increases, profit increases and the stagnation or reduction of wages. This 'social mix' can be a recipe for a future social storm. With such a context, it will only take a small 'spark' to ignite this social storm. However, this social mix is not just a Western phenomenon. There are many indications that the Chinese economy is exposed to the same social forces regarding insecure jobs, rising prices, rising profits and wage stagnation. In China, this social mix is just as dangerous for social stability as it is in the West because approximately 700 million people in the Chinese middle class are constantly seeking to meet their expectations and improve their standard of living. The Chinese leadership is aware that if they can't meet these expectations, there may arise a social demand for political renewal (Johannessen, 2021, 2022a).

The economic goal of political parties during the post-war period was full employment (or low unemployment); this was also an important political goal during the crises of the 1920s and 1930s. In other words, economists and politicians have used unemployment indicators to provide information about whether or not society is moving in the right direction. However, as we move towards the Fourth Industrial Revolution, it seems reasonable to say that unemployment indicators no longer provide reliable information. On the contrary, unemployment statistics seem to function more as a veil obscuring what is happening rather than giving information about the state of society. We have pointed this out several times when we say that the effects of automation resulting from the deployment of artificial intelligence and intelligent robots are not visible in the unemployment statistics but in the wage statistics. Consequently, standard economic indicators have become less informative as we move towards the Fourth Industrial Revolution. If this economic indicator is no longer reliable, what other factors should we use that will provide more reliable information? Today prices and profits are high, while wages fail to keep up. In the short version, a new social structure is developing, which we have termed elsewhere feudal capitalism (Johannessen, 2023). The development of feudal structures is also described and analysed by Kotkin (2020). He points out that the middle class will be the most negatively affected by such a neo-feudalist development, especially those who do not have skills compatible with the new technology. In addition, the working class will also be negatively affected by this social revolution.

In the West, large sections of the working and middle classes felt side-lined during the first phase of globalization when the Chinese economy benefitted from producing cheap consumer goods. These people became disillusioned, powerless and alienated from what they saw developing in the workplaces, which, for generations, had secured their jobs in their local areas. In the US, this created fertile ground for the growth of new nationalist and protectionist political movements, the consequences of which are now becoming evident. The reinforcement of this trend may contribute to growing feudal, capitalist structures in both the US and Europe.

The pace of change outstrips our ability to adapt

If the organizational leadership introduces too many consecutive changes, this may unintentionally require people to change much too quickly. This situation can result in erratic organizational behaviour. Consequently, this can lead to resistance to change. This situation may alienate those who initially supported the necessity of change and give more weight to those opposed to change.

The leadership can prevent such erratic behaviour by involving employees at an early stage and in planning of changes. In the planning phase, they should present the change project as a win-win situation where employees can make substantial gains while they risk losing little. In this way, everyone knows what must be done, why it should be done, how it should be done and the desired effects of the changes.

When changes occur faster than people can adapt, it seems reasonable to assume that resistance to change will increase. This situation is where the Prospect Theory of Nobel laureate Kahneman (Kahneman & Tversky, 1979, 2000) can be of help in understanding both what happens when changes happen quickly, and what can be done at the individual, organizational and societal levels, to mitigate resistance to change and enable people to adapt to changes.

The problem we examine here is people's resistance to organizational change. We will address the question: How can we use Prospect Theory to explain why people resist organizational change?

The aim is to identify how leaders can reduce resistance to change and identify explanations of why people resist organizational change. The point here is how people relate to risks. Risk relates to our assumptions about potential outcomes and how decision-makers evaluate these outcomes.

Kahneman and Tversky developed the Prospect Theory in 1979.[16] The theory holds that when people face a risk situation with limited information and do not apply rigorous analytical processes, their choices will often be driven by how they or others frame the information (Wolfe, 2008: 6).

The core of Prospect Theory consists of the assessments people make with what they may gain or lose in making a choice. One example of such a choice might be whether or not to engage actively in a change process within an organization. According to Prospect Theory, the possibility of losing an existing position will generate a level of resistance that will outweigh the energy and resources a person might expend to gain a new position (Kahneman, 2011: 279–280). Most people are averse to losing something that they already have.

Our assessments are largely biased, distorted, and not wholly reliable. Regardless, we make considerable use of these assessments in decision-making. Tversky and Kahneman's research demonstrated that these assessments are often based on heuristics or 'rules of thumb', which people use in decision-making (Tversky & Kahneman, 1974: 1124–1131). A basic assumption of Prospect Theory is that people use these rules of thumb without even realizing they are doing so.

In Chapter 5, we will elaborate on various aspects of Prospect Theory. We will briefly describe three aspects of Prospect Theory here, which we will analyse and reflect upon thoroughly in Chapter 5. The first aspect is 'decision-making under uncertainty'. The second aspect is how we frame a situation and how people have a fundamental aversion to risk. The third aspect is how we apply heuristic assessments. In particular, we will look at how the availability and use of anchoring govern our assessments and decisions.

Decision-making under uncertainty

At first, it may seem reasonable to assume that people will seek out risk if they live in poor conditions. The idea is that the situation can't get worse, so people take risks to improve their lives. According to Prospect Theory, however, this intuitive assumption is incorrect. On the contrary, when a person faces the possibility of losing the rights, power, positions, income, etc., already achieved, he or she will seek to retain what they have gained and be reluctant to change; one may imagine people avoiding participating in change processes for as long as possible because they risk losing what they have achieved (Kahneman & Tversky, 2000: 22). One may envisage people avoiding participating in change processes for as long as possible because they risk losing what they have achieved.

In Prospect Theory, the 'certainty effect' (Kahneman & Tversky, 2000: 17) explains why people are risk-averse. In broad terms, this effect is the preference for the certain over the possible.

Framing

One assumption in Prospect Theory is that people do not have complete information when making decisions. They almost always act based on the data at hand. From this, the theory does not assume people are rational when making choices. The theory investigates how people act in practice when making choices, asking, for example, how they use intuition when making choices in uncertain situations. When faced with a choice between uncertainty that may offer future opportunities and a current status quo situation, people often act based on the proverb, 'A bird in the hand is worth two in the bush'. In other words, we tend to choose the safe option over the one that is uncertain but offers opportunities.

Heuristic assessments

There are four basic heuristic assessments that Tversky and Kahneman have described (Kahneman, 2011: 109–269; Tversky & Kahneman, 1974). These are:

1. Representativeness and randomness
2. Anchoring
3. Availability
4. Validity

Notes

1. https://en.wikipedia.org/wiki/Open_Letter_on_Artificial_Intelligence
2. https://en.wikipedia.org/wiki/Open_Letter_on_Artificial_Intelligence
3. We make a distinction here between the first three industrial revolutions: the First Industrial Revolution around 1760 (the steam engine); the Second Industrial Revolution around 1870 (science, electricity and mass production); and the Third Industrial Revolution around the 1950s (the rise of digital technology). The Fourth Industrial Revolution will be characterized by robotics, artificial intelligence and related technologies.
4. https://en.wikipedia.org/wiki/Open_Letter_on_Artificial_Intelligence
5. Brynjolfsson, E. & Saunders, 2013; Bessen, 2020; Chesbrough, 2003; Christensen, 2015; Ford, 2016; Frey, 2019, 2020; Leontief, 1979, 1986; Perez, 2003; Peters et al., 2019; Shanahan, 2015; Susskind, 2020; Webb, 2021, 2022;
6. See Appendix 3.
7. https://en.wikipedia.org/wiki/Herodotus
8. Acemoglu et al., 2015; Atkinson, 2015; Picket, 2017; Piketty, 2014, 2016; Stilwell, 2019.
9. (Li, 2021; Miao, 2019; Rocca, 2018; Johannessen, 2021, 2022b)
10. Acemoglu & Robinson, 2020; Rachman, 2022; Guriev & Treisman, 2022; Kotkin, 2020.
11. https://en.wikipedia.org/wiki/AlphaZero
12. https://en.wikipedia.org/wiki/Stockfish_(chess)
13. https://en.wikipedia.org/wiki/Janus
14. https://en.wikipedia.org/wiki/Tiananmen_Square
15. STEM stands for Science, Technology, Engineering and Mathematics.
16. Kahneman & Tversky, 1979: 263–292. Kahneman won the Nobel Prize for Economics in 2002 for his work with Amos Tversky on decision-making under uncertainty, including, amongst other things, prospect theory.

References

Acemoglu, D. & Autor, D.H. (2011). Skills, tasks and technologies: Implications for employment and earnings, in Card, D. & Ashenfelter, O. (eds.), *Handbook of labor economics*, vol. 4, Elsevier, Amsterdam, pp. 1043–1171.

Acemoglu, D. & Robinson, J.A. (2020). *The narrow corridor: How nations struggle for liberty*, Penguin, New York.

Acemoglu, D.; Suresh, N.; Pascual, R. & Robinson, J.A. (2015). Democratic redistribution, and inequality (Chapter 21), in Atkinson, A.B. & Bourguignon, F. (eds.), *Handbook of income distribution*, vol. 2, Elsevier, London, pp 1885–1966.

Adriaenssen, D.J. & Johannessen, J-A. (2016). Conceptual generalizations, Methodological reflections in social science: A systemic viewpoint, *Kybernetes*, 44 (4): 588–605.

Adriaenssen, D.J. & Johannessen, J-A. (2016a). Prospect theory as an explanation for resistance to organizational change: Some managerial implications, *Problems and Perspectives in Management*, 14 (2): 84–92.

Aghion, P.; Antonin, C. & Bunel, S. (2021). *The power of creative destruction*, Harvard University Press, Cambridge, Massachusetts.

Ainley, P. (2014). Follow your dreams and attend to universities if possible, *Latitude*, 21 December.

Ainley, P. (2016). *Betraying a generation: How education is failing Young people*, Policy Press, Bristol.

Allen, R.C. (2009). *The British industrial revolution in global perspective*, Cambridge University Press, Cambridge.

Allison, G. (2018). *Destined for war: Can America and China escape Thucydides trap?* Scribe, London.
Atkinson, A.B. (2015). *Inequality: What can be done*, Harvard University Press, Cambridge.
Atkinson, R.D. & Ezell, S.J. (2014). *Innovation economics: The race for global advantage*, Yale University Press, New York.
Autor, D. (2015). Why are there still so many jobs? The history and future of workplace automation, *Journal of Economic Perspectives*, 29 (3): 3–30.
Bessen, J. (2020). Attitudes to technology: Part 1, in Skidelsky, R. & Craig, N. (eds.), *Work in the future*, Palgrave, London, pp. 83–88.
Bruder, J. (2017). *Nomadland: Surviving America in the twenty-first century*, W.W.Norton, New York.
Brynjolfsson, E. & Saunders, A. (2013). *Wired for innovation: How information technology is reshaping the economy*, The MIT Press, London.
Bunge, M. (1974). *Sense and reference*, Reidel, Dordrecht.
Bunge, M. (1996). *Finding philosophy in social science*, Yale University Press, New Haven.
Bunge, M. (1998). *Social science under debate*, University of Toronto Press, Toronto.
Bunge, M. (1999). *The sociology-philosophy connection*, Transaction, New Brunswick, NJ.
Carey, N. (2019). *Hacking the code of life: How gene editing will rewrite our futures*, Icon Books, New York.
Carney, T.P. (2020). *Alienated America*, Harper, New York.
Chesbrough, H.W. (2003). *Open innovation: The new imperative for creating and profiting from technology*, Harvard Business School Press, Boston.
Chetty, R.D; Grusky, M.; Hell, N.; Hendren, R. Manduca, R. & Narang, J. (2017). The fading American dream: Trends in absolute income mobility since 1940, *Science* 356 (6336): 398–406.
Christensen, C.M. (2015). *The innovators dilemma: When new technologies causa great firms to fail*, Harvard Business Review Press, Boston.
Eubanks, V. (2017). *Automating inequality*, St. Martin's Press, New York.
Ford, M. (2016). *The rise of the robotics, technology and the threat of mass unemployment*, Oneworld, New York.
Frank, T. (2004). *What's the matter with Kansas? How conservatives won the Heart of America*, Metropolitan Press, New York.
Frank, T. (2020). *People without power*, Scribe, London.
Frey, C.B. (2019). *The technology trap: Capital, labor, and power in the age of automation*, Princeton University Press, Princeton.
Frey, C.B. (2020). Attitudes towards technology: Part II, in Skidelsky, R. & Craig, N. (eds.), *Work in the future: The automation revolution*, Palgrave, London, pp.89–97.
Gaskell, P. (1833). *The manufacturing population of England, its moral, social and physical conditions*, Baldwin and Cradock, London.
Guriev, S. & Treisman, D. (2022). *Spin dictatorship: The changing face of tyranny in the 21*st *century*, Princeton University Press, New York.
Johannessen, J-A. (2018). *Innovation leads to economic crises*, Palgrave, London.
Johannessen, J-A. (2020). *The workplace of the future*, Routledge, London.
Johannessen, J-A. (2020a). *Automation, innovation and economic crises: Survival the fourth industrial revolution*, Routledge, London.
Johannessen, J-A. (2020b). *Artificial intelligence, automation and the future of competence at work*, Routledge, London.
Johannessen, J-A. (2021). *China's innovation economy: Artificial Intelligence and the new silk road*, Routledge, London.
Johannessen, J-A. (2021a). *Artificial intelligence, automation and ethics in the innovation economy*, Routledge, London.
Johannessen, J-A. (2021b). *Communication as social theory: The social side of knowledge management*, Emerald, London.

Johannessen, J-A. (2022). *Creativity, innovation and the fourth industrial revolution: The da Vinci strategy*, Routledge, London.
Johannessen, J-A. (2022a). *The new silk road and the innovation economy in China*, Routledge, London.
Johannessen, J-A. (2022b). *The philosophy of tacit knowledge*, Emerald, London.
Johannessen, J-A. (2022c). *A systemic approach to continuous change in the innovation economy*, Routledge, London.
Johannessen, J-A. (2022d). *Intelligent robots, consciousness and creativity: The search for hidden knowledge, The cognitive side of knowledge management*, Emerald, London.
Johannessen, J-A. (2023). *Feudal capitalism: The political economy of the innovation society*, Routledge, London.
Johannessen, J-A. (2024). *Safe jobs in the innovation economy: Strategic competence creation*, Routledge, London.
Johannessen, J-A. (2024a). *Uncertain jobs in the innovation economy: Strategic competence destruction*, Routledge, London.
Kahneman, D. (2011). *Thinking fast and slow*, New York, Allen Lane.
Tversky, A., & Kahneman, D. (1974). Judgment under uncertainty: Heuristics and biases. *Science*, 185, 1124–1131.
Kahneman, D. & Tversky, A. (1979). An analysis of decision under risk, *Econometrica, Journal of the Econometric Society*, 47 (2): 263–292.
Kahneman, D. & Tversky, A. (2000). Prospect theory: An analysis of decision under risk, in Kahneman, D. & Tversky, A. (eds.), *Choices, values and frames*, Cambridge University Press, Cambridge, pp. 17–43.
Kessler, S. (2018). *Gigged: The end of the job and the future of work*, Macmillan, New York.
Kirzner, S. (1982). The theory of entrepreneurship in economic growth. In Kent, C.A.; Sexton, D.L. & Vesper, K.H. (eds.), *Encyclopaedia of entrepreneurship*, Prentice Hall, Englewood Cliffs, NJ.
Kochan, T.A. & Dyer, L. (2021). *Shaping the future of work: A handbook for action and a new social contract*, Routledge, London.
Kotkin, J. (2020). *The coming of neo-feudalism: A warning to the global middle class*, Encounter Books, New York.
Kurzweil, R. (2008). *The age of spiritual machines: When computers exceed human intelligence*, Penguin, London.
Leontief, W. (1979). Is technological unemployment inevitable? *Challenge*, 22 (4): 48–50.
Leontief, W. (1986). *The future impact of automation on workers*, Oxford University Press, Oxford.
Li, C. (2021). *Middle class Shanghai,: Reshaping US-China engagement*, Brooking Institution Press, Washington, DC.
Mantoux, P. (1961). *The industrial revolution in the eighteenth century: An outline of the beginning of the modern factory system in England*, Routledge, London.
McGarvey, D. (2018). *Poverty safari*, Picador, London.
Miao, Y. (2019). *Being middle class in China: Identity, attitudes and behaviours*, Routledge, London.
Milanovic, B. (2016). *Global inequality*, The Belknap Press, New York.
Perez, C. (2003). *Technological revolutions and financial capital*, Edward Elgar, London.
Peters, M.A.; Jandric, P. & Means, A.J. (2019). *Education and technological unemployment*, Springer, London.
Petras, J. & Veltmeyr, H. (2011). *Beyond neoliberalism: A word to win*, Routledge, London.
Picket, K. (2017). Foreword, in Brown, R. (ed.), *The inequality crises*, Policy Press, London, pp. vii–viii.
Piketty, T. (2014). *Capital in the twenty-first century*, The Belknap Press of Harvard University Press, Boston.

Piketty, T. (2016). *Chronicles: On our troubled times*, Viking, London.
Polanyi, M. (1966). *The tacit dimension*, Doubleday, New York.
Polanyi, M. (1969). *Knowing and being*, University of Chicago Press, Chicago.
Polanyi, M. (2015). *Personal knowledge*, University of Chicago Press, Chicago.
Rachman, G. (2022). *The age of the strongman: How the cult of the leader threatens democracy around the world*, Bodley Head, New York.
Reeves, R.V. (2018). *Dream Hoarders: How the American upper middle class is leaving everyone else in the dust. Why that is a problem and what to do about it*, Brooking Institution Press, Washington, DC.
Rocca, J-L. (2018). *The making of the Chinese middle class: Small comfort and great expectations*, Palgrave, London.
Schwab, K. (2016). *The fourth industrial revolution*, World Economic Forum, Geneva.
Schwab, K. (2018). *Shaping the fourth industrial revolution*, World Economic Forum, Geneva.
Seligman, M.E. (2006). *Learned optimism*, Vintage Books, New York.
Seligman, M.E. (2011). *Flourish: A new understanding of happiness and well-being*, Nicolas Brealey, New York.
Shanahan, M. (2015). *The technological singularity*, MIT Press, Boston.
Shipler, D. (2005). *The working poor*, Vintage, New York.
Simon, H.A. (1965). *The shape of automation for men and management*, New York: Harper & Row.
Simon, H.A. (2019). *The sciences of the artificial: Reissue of the third edition with a new introduction by John Laird*, MIT Press, Boston.
Skidelsky, R.; Craig, N. (eds.) (2020). Introduction, in *Work in the future: The automation revolution*, Palgrave, London, pp 1–7.
Slobodian, Q. (2020). *Globalists: The end of empire and the birth of Neoliberalism*, Harvard University Press, Boston.
Standing, G. (2014). *The precariat: The new dangerous class*, Bloomsbury Academic, New York.
Standing, G. (2014a). *A precariat charter*, Bloomsbury, London.
Standing, G. (2016). *The corruption of capitalism: Why rentiers thrive and work does not pay*, Biteback Publishing, London.
Standing, G. (2019). *Plunder of the commons, A manifesto for sharing public wealth*, Pelican, New York.
Stilwell; F. (2019). *The political economy of inequality*, Polity, London.
Susskind, D. (2020). *A world without work: Technology, automation and how we should respond*, Allen Lane, London.
Toffler, A. (1970). *Future shock*, Bantam Books, New York
Watzlawick, P. (1980). *Ultra-solutions: How to fail most successfully*, W.W. Norton, New York.
Webb, A. (2021). The future today institute's 14th annual tech trends report. https://futuretodayinstitute.com/trends/
Webb, A. (2022). *The genesis machine: Our quest to rewrite life in the age of synthetic biology*, Public Affairs, New York.
Wolfe, W.M. (2008). *Winning the war of words*, Praeger, London.
Woodcock, J. & Graham, M. (2019). *The gig economy: A critical introduction*, Polity, London.

2 Automation is here to stay

Core idea in this chapter

A significant source of tension in relation to automation and new technology is the difference between short- and long-term perspectives.

Key points in this chapter

- Inflation, rising profits and stagnant wages are all the consequences of automation caused by the introduction of new technology.
- In this new wave of automation, five kinds of competence will be crucial for success in finding highly paid jobs.
- The automation that is the result of artificial intelligence and intelligent robots is creating simultaneous economic growth and inequality.
- The automation that results from artificial intelligence will trigger innovations in five areas.
- The new wave of automation will promote the development of new methods of leadership (6-C leadership).
- Automation will be implemented in settings where real and relative productivity are suffering the largest declines.
- In the period 2035–2040, half of all jobs in the US will be taken over by intelligent robots.

Introduction

In this chapter, we will examine the following question: What is the potential impact of automation on future employment trends?

Our goal is to develop ideas that may be of practical benefit. Such ideas must be based on facts, scholarship and reflections on the application of new technology. The ideas we develop throughout this book are intended to be future-proof, in other words, they should continue to be relevant in a future where workplaces have been transformed by new technology. We are clear that the Fourth Industrial Revolution will lead to the emergence of many new technologies. In this book, however, we examine what we believe to be the fundamental technologies driving the onset of the Fourth Industrial Revolution: Artificial intelligence and intelligent robots.

DOI: 10.4324/9781003567325-2

Automation is nothing new. Why then should we examine future developments in automation processes, when such processes have been a feature of the economy ever since the First Industrial Revolution began around 1760 in Britain? The answer is straightforward. Future automation processes will affect everyone, absolutely everyone! Automation processes and machines in the 18th and 19th centuries mainly replaced human physical labour. Automation processes in the late 20th century greatly replaced administrative work functions. However, in the future, advances in artificial intelligence (AI), robotics and related technologies will most probably facilitate the automation of many of the work functions of information workers, knowledge workers, middle managers and others. Schwab (2016, 2018) terms this future technological development the Fourth Industrial Revolution.

As mentioned, AI will constitute the prime technology of the Fourth Industrial Revolution. However, there are also many other related new technologies that will sustain and drive the Fourth Industrial Revolution. This is why we can also term this industrial revolution the innovation economy. This economy is constantly being driven forward by new innovations. There are so many innovations that have already emerged that it is perhaps more correct to say that there are cascades of innovations that are affecting people, organizations and society. These innovations are at different levels and have different impacts on workplaces. Some of these innovations, especially those related to AI and robotics, are changing people's lives and workplaces. Other innovations affect people's lives to a lesser extent, but are making an impact at an overarching level, for example, in the competition between China and the US (Johannessen, 2021, 2022). In other words, the Fourth Industrial Revolution will also be characterized by competition between the two largest global economies (Johannessen, 2023a). In an initial phase, this will focus on gaining control of the microchips production and market. Microchips are absolutely essential to modern production methods and are used in various technologies, such as AI and robotics and in numerous devices, such as computers, smartphones and countless other electronic devices.

Allegorically, we might say that when this technological revolution collides with the future, it will create consequences for us all. This will first become evident in increasing inflation and corporate profits, while wages will fall out of step with rising inflation (Aghion et al., 2021). All the indicators seem to point in the direction of an inflation-driven economy, where many people will suffer. One of the reasons for this development is that globalization, which focused on low-cost countries, such as China, supplying cheap consumer goods to the West, and Western capital being invested in low-cost countries, is undergoing change. Production that was previously outsourced to China is slowly but surely being moved back to the West due to increasing tensions between China and the US (Allison, 2018; Johannessen, 2023a). In the West, wages are higher. Thus, the costs of producing are also higher. The US and Western countries have banned some Chinese high-tech companies from Western markets, thus reducing competition, while stating that it has been done for national security reasons. China is focusing on production for the domestic market to a greater extent. The domestic market in China represents a great potential for Chinese companies, because the Chinese middle class numbers approximate 700 million people.[1] In addition, as of 1 January 2022, China entered into the Regional Comprehensive Economic Partnership (RCEP) Agreement, which represents the world's largest free trade area, encompassing approximately 30 per cent of the world's population and approximately 30 per cent of all global production. Moreover, China is increasing its involvement in the production and distribution of infrastructure projects along the New Silk Road (Johannessen, 2022).

The fact that the Fourth Industrial Revolution is coinciding in time with the increasing competition between the West and China is not accidental. It can be explained by 'Thucydides Trap', which proposes that a conflict can develop when a rising great power (e.g., China) challenges the dominant great power (e.g., the US). The metaphor, 'Thucydides' Trap', has been used by Allison in his book, *Destined for war: Can America and China escape Thucydides Trap?* (2018). Thucydides (2009) was an Athenian historian who wrote about the war between Sparta and Athens in which the two powers were battling for hegemony in the classical Greek world. To prevent the further rise of Athens, the Spartans went to war with the Athenians. According to Thucydides, the war between Sparta and Athens was inevitable due to the fact that the dominant power (Sparta) was challenged by the rising power (Athens). Through a major research project at Harvard, Allison (2018) has examined 16 conflicts between rising and dominant powers over a period of 500 years and concluded that 12 of these led to war. Therefore, the Fourth Industrial Revolution and the competition we see developing between China and the US, will, most probably, end in a confrontation that is not only based on technological competition.

Artificial intelligence is such a powerful technology that all of today's work processes will be changed by the time 'the robot shock' has fully impacted society during the Fourth Industrial Revolution. However, in this context, the word 'revolution' is a misnomer, because it will in reality take many decades before the new technology has become an integral part of people's everyday lives (Schwab, 2016, 2016). However, when this technology has been fully adopted, our workplaces will become unrecognizable. This is evident if we consider the developments in the three previous industrial revolutions. Of course, the Fourth Industrial Revolution may be different, but no evidence seems to indicate this. If the above is correct, then most people will have enough time to adjust to the upcoming 'robot shock'.

It is not only people in their workplaces who will be affected by the new technology. Organizations will have to search for employees with qualitatively new competences. Although it will take a long time before this technology is implemented in most work processes, there will be some industries and businesses that will be affected by this robot shock long before others.

We have pointed out in several places how this technology can influence the development of automation processes (Johannessen, 2020a, 2020b). This time we will examine how robotization will affect our workplaces in the short and long term and which strategies we can develop to survive the future robot shock.

Artificial intelligence and intelligent robots will first change work processes where:[2]

1. The real and relative costs are high.
2. The real and relative productivity is falling.
3. The real and relative quality is falling.
4. The rate of spread of innovations is greatest.
5. New knowledge has the possibility of being developed into new technology.

Some of the sectors that will first be affected by new innovations in relation to the above points will be within pharmacy and medicine; the military; and biology research (Ford, 2021). However, increased automation is only one aspect of the future applications of AI and robotics. Another aspect that is linked, but distinct, is how the new technology will also result in new organizational models being developed.[3] One of the consequences of these developments will be a further fragmentation of the middle class, which we have already witnessed in the West in general and the US in particular (Reeves, 2018). However, this fragmentation of the middle class has not yet occurred in China. On the contrary, the Chinese middle class has grown very strongly in the last 30 years, from the 1990s to the 2020s. If the same fragmentation of the middle class were to occur in China, then it can be expected that the robot shock will lead to major tensions in the Chinese system. The tensions could become so great that the political system in China could experience major difficulties. The Chinese Communist Party survives only as long as it can meet the expectations of the Chinese middle class (Mahbubani, 2018, 2020). If the introduction of new technology should negatively affect the expectations of the Chinese middle class, then we will with a great degree of certainty see changes in the Chinese political system (Fang & Fu, 2020; Mattingly, 2020; Joseph, 2019). Should the US have a grand strategy to change China's political system, it lies precisely here, in giving China access to all the imaginable new technology related to AI and robotics, because such a development would lead to roughly the same fragmentation of the middle class in China that we have seen in the US.

Information systems, cybernetic systems, systemic thinking, expert systems, cognitive technologies, automatic decision-making machines, algorithms, learning algorithms, etc., have been part of the prelude to today's practical artificial intelligence tools. It can be said that AI, robotics and cybernetics have had an evolutionary historical development all the way back to the 1950s with pioneers in the field, such as Norbert Wiener, Ross Ashby, Stafford Beer, Gregory Bateson, Ernst von Glasefeld, Humberto Maturana and Gordon Pask.[4] In other words, although AI and robotics will most probably form the core technology of the Fourth Industrial Revolution (Schwab, 2016), the technology has a long historical evolutionary development behind it.[5]

This technology will utilize huge volumes of data (Big Data), which, in combination with algorithms, will enable the identification of patterns and trends and decisions related to human behaviour. For example, artificial intelligence and Big Data are being used in China to manage road traffic. However, such monitoring systems can be used for other things than just managing and controlling traffic. This large amount of data can be used for almost anything related to human behaviour.

Companies such as Google, Amazon, Facebook, Instagram and so on have control over large amounts of data. When this data is used in combination with artificial intelligence, the above companies are able to transform data into information and knowledge about their customers' behaviour. With this knowledge, the companies and their Chinese counterparts can predict how customers will react to trends and changes, long before the customers know themselves. It is, amongst other things, in this area that artificial intelligence can get out of control and take control of people's behaviour and lives.

When, not if, this knowledge is used to automate our behaviour, then it can be said that artificial intelligence has taken control of our lives. Our behaviour will have been automated in the same way that our workplaces are automated by using this technology.

When artificial intelligence and intelligent robots are utilized to increase productivity, many service jobs will be created around this technology. We have seen this development in the delivery of goods and services in the big cities, in part-time jobs in supermarkets and the increase in unskilled jobs in the service industry. In other words, the increased use of this new technology is creating very large wage differences. This can be explained by the historical fact that higher wages are closely linked to those who have skills related to new technology. In this way, the new technology will result in an even greater economic inequality than at present. The utilization of the technology will simultaneously increase economic growth and economic inequality, in addition to changing work processes in revolutionary ways. It is this development that we will develop survival strategies for, at the individual, organizational and social levels.

We have previously examined which jobs will be secure in the innovation economy (Johannessen, 2024). In the aforementioned study, we found five areas where the probability of secure jobs was high. These jobs can be linked to the following areas:

1. High-tech competence, i.e., those people whose competence is compatible with the new technology.
2. Dual competence, i.e., those professionals, such as psychologists, that not only have traditional expertise within the field but also competence related to AI that they can utilize to increase productivity. In concrete terms, this can be skills in coding and the use of algorithms that will enable them to develop personalized programmes.
3. The traditional knowledge worker. Those people with an extensive higher education that are needed in order to make sure the wheels of the education system go round.
4. The creative innovation worker. These are the creative people that are able to generate innovations.
5. The entrepreneur. These are the people who set up businesses on the basis of their own expertise in the new technology, as well as utilizing the skills of others.

Narratives related to automation

Case letter 1: 6-C leadership

Artificial Intelligence and Emerging Technology (AIET) is concerned with developing various technologies based on artificial intelligence in order to make systems for management and control more efficient. Most management systems in social hierarchies have grown strongly and are very labour-intensive. However, with the new technology, it will be possible to automate management systems to a certain degree. The digitization and informatization that are developing in these management structures will increase productivity. As we move towards the Fourth Industrial Revolution, many theorists believe that management systems in organizations and institutions will to a large extent become automated sometime around 2035–2040 (Kurzweil, 2005, 2008, 2013).

It is the centralized structures that will be the target of these operations. Productivity decreases when those who do what the system is designed to do become fewer, due to the fact that many are employed in administrative and management systems. When this occurs, the real and relative productivity is reduced. This development has been evident over the last 30 to 40 years, where systems have become increasingly large, with increasing centralization. For example, this concerns large systems such as universities, hospitals, public institutions, banks, financial institutions, manufacturing companies and so on. According to innovation theory and practice, it is under such conditions that automation will be introduced in order to counteract the fall in productivity.

This process will initially be implemented using new technology to automate established work processes, such as an extensive digitization of administrative systems. Subsequently, it is reasonable to assume that the organization of the work processes will be changed. In a third phase, fewer people will be required for these functions. The new technology that will increasingly be introduced, and to a certain extent has already been introduced through various digitization and informatization technologies, will change what can be called the C-6 system, that is, command, control, communication, coordination, cooperation and co-creation. These six functions will utilize specific technologies in order to increase efficiency.

The 6-C system is related to the central management functions in organizations, institutions and other social systems. In the future, it is highly probable that command structures will be automated by the new technology – this is already occurring to some extent. The new technology can be utilized to send and receive data and information more efficiently between the various operations in organizations. The same development will occur in the control systems; in how we communicate with each other; in the way we coordinate our work tasks; in how we collaborate; and last, but not least, how we co-create innovations.

A simple proposition states that where functions, activities or processes experience falling productivity, then innovations will emerge to improve productivity. This is the basis for the assumption that the 6-C system will mainly be affected in the labour-intensive administrative and management systems, which have developed in the neoliberal era since the 1980s. In other words, it has taken at least 40 years before people have begun to focus on the fact that the over-bureaucratization of the administrative and management sectors of organizations are hindering productivity.

The consequences of the above development have been higher costs and lower flexibility. The rationale is that all parts of the administrative and management systems need to be involved in the processes that lead to decisions. Moreover, people rarely quit their jobs voluntarily and often present arguments why they should retain their jobs. In other words, there is often great resistance to any changes that can lead to the automation of people's work tasks. When the people whose work tasks need to be automated in order to increase productivity are the same people who also have decision-making authority, both directly and relationally in an organization, then this can partly explain why it takes so long to introduce automation processes in those parts of organizations concerned with leadership and management.

Description related to automation

Automation concerns the use of technologies that can reduce human intervention in work processes.[6] Historically, automation processes can be traced as far back as to the Ancient Greeks, approximately 300 BCE, and to the Arabs, approximately 13th century AD (Guarnieri, 2010: 42–43). However, it was not until the First Industrial Revolution in Britain, from around 1750, that automation was first used to greatly increase productivity and economic growth (Bennett, 1979: 47). There are many advantages and drawbacks associated with automation (Autor, 2015: 3–30). Autor describes some of the advantages: Increased productivity, improved quality in production, increased predictability in the work processes, greater degree of robustness in the production processes, reduced use of human labour in the work processes, reduced time from the start of a process to its completion, less repetitive work tasks for workers, as well as increased freedom for workers to do things other than manual labour.

In this chapter, we will examine the following question: What is the potential impact of automation on future employment trends? Regarding this question, there are various aspects of automation that may affect the workplace of the future. These aspects are not only linked but also distinct (Author, 2015). The new technology that can affect the workplaces of the future, that is, AI and robotics, is already being used in many work processes, although it is still at a low level of application. Robotics, in conjunction with digitization and informatization, is being used to automate work processes in businesses, institutions and public administration. Automation in general, and robotics in particular, are being utilized to increase production and economic growth. However, this economic growth is also creating greater economic inequality. In conclusion, we will, in the following, discuss how five aspects of automation, together and individually, will affect changes in our workplaces. The aspects we will describe are the following: digitization and informatization; intelligent robots; economic growth; economic inequality; and changing work processes.

Digitization and informatization

The above description provides a backdrop for assessing the upcoming robot shock, the future automation at different levels of business and industry, and the different effects on workplaces. In recent times, automation took off in earnest with the implementation of digitization and informatization processes in the 1990s. Amongst other things, digitization concerns the interconnection of computers and the internet, and the process of converting information into digital formats. In addition, the continuous development of smart phones has greatly facilitated communication and information between people and businesses. Informatization in this context refers to the extent by which workplaces have become information-based; in other words, the fact that new communication technologies are used as a means for furthering productivity. Computers and smartphones have gradually started to utilize artificial intelligence affecting the ways in which work processes are carried out. For a long time, however, the organization of workplaces was not to a large extent influenced by the new technology. COVID-19 triggered the potential of the new technology in relation to a re-organization of workplaces. Many organizations shut their doors, and business was conducted remotely. In other words, people started working remotely from their home offices. This new organization focused on the productivity of workers. It was what people produced, how they produced it and the efficiency of production that became important, not the location where people did their work. We can term this new re-organization of work, 'robot organisation'. In practice, this meant that the role of managers and leaders was completely changed. In other words, this new type of leadership of workers was not the physical management of people located in their places of work. In such a new situation, the focus was on results and the performance of individual workers. Many leaders and workers found it difficult to cope with this new situation, and in many instances leaders tried to 'force' people back to their workplaces in the hope that they could gain a greater overview of what individual workers were doing. In this context, the robot shock was also a 'shock' for many leaders who could not cope with having to lead people they rarely or never saw.

Digitization and informatization processes have had a common denominator, which is that white-collar jobs have been subjected to the largest changes. Many blue-collar jobs were negatively affected by automation and its consequences during the first two industrial revolutions. However, during the Third Industrial Revolution, the digitization and informatization of work processes greatly impacted many white-collar jobs. In addition, during the Third Industrial Revolution and at the beginning of the Fourth Industrial Revolution, the middle class in Western countries came under pressure (Reeves, 2018). Many middle-class workers had to take poorly paid and insecure jobs in the service sector (Standing, 2014, 2014a, 2016, 2019). In China, on the other hand, digitization and informatization processes favoured the middle class. By 2022, the Chinese middle class had grown considerably and was now around 700 million people (Johannessen, 2021, 2022). However, it was not only digitization and informatization processes that led to the growth of China's middle class, although it had some significance. The most important factor was perhaps the strong investment in competence development, from the 1980s onwards. Regardless of the reasons for the large growth in China's middle class, digitization and informatization processes had different consequences for the West and China and also other Asian countries.

A major study conducted at Oxford University in England has concluded that in the period 2035–2040, half of all jobs in the US will be taken over by digitization, artificial intelligence and intelligent robots (Oppenheimer, 2019: 3). Thus, the study indicates that between 2035–2040, the consequences of the robotization of workplaces will have become a reality. In this context, the learning of new skills will take a long time. However, it seems clear that in order to survive the robot shock, a comprehensive plan for the reorganizing of education, from primary to higher education, will be needed, so people and organizations can acquire new skills. Ainley has addressed this issue in his article, 'Follow your dreams and attend to universities if possible' (2014) and his book, *Betraying a Generation: How Education is Failing Young People* (2016). He shows, in an exemplary fashion, the non-validity of the idea that if young people complete a traditional higher education, they will be able to find a secure and well-paid job. This implies that young people should rather acquire skills that are compatible with the new technology.[7]

Intelligent robots

Historically, new technology and automation has focused on increasing productivity. This occurred during the first three industrial revolutions: The First Industrial Revolution (steam power); the Second Industrial Revolution (electrification); and the Third Industrial Revolution (digital technology). The Fourth Industrial Revolution will most probably be characterized by the implementation of new technologies, such as AI, robotics and related technologies. In other words, the implementation of these new technologies will lead to increases in productivity and also to a re-structuring of workplaces and working life. It is precisely this future scenario which this book focuses on: How will people be able to survive the 'robot shock'? That is, the restructuring of working life and the demand for new skills that will result from the future applications of AI, robotics and related technologies.

Some people will profit greatly from the economic and social disruption that the robots will have on society. Others who fail to adapt to the increasing changes resulting from the introduction of the new technology will very likely suffer. Throughout history, there are three main strategies that people have adopted to deal with the introduction of new revolutionary technology in workplaces. The first and successful strategy is those people that quickly adapt to the new technology by acquiring compatible new skills. These people will be the winners in the time to come. The second strategy is those people who resist change and do everything to hinder the introduction of the new technology. For instance, during the First Industrial Revolution, the Luddite weavers smashed the new textile machines, because they lost their jobs due to the new technology.[8] Historically, people who attempt to hinder the introduction of new technology have lost out in the long run. The third strategy is those people who, instead of adapting to the new technology, find new workplaces that have not been immediately impacted by the introduction of new technology. These people will manage to get by for a long time, but their jobs will not be able to maintain the same salary level as jobs in companies that have invested in new technology, because productivity will be lower. In other words, they will eventually sink to the bottom of the wages hierarchy, although it may take some time for this to happen.

Since the First Industrial Revolution in Britain around 1750, the Western world has had an underlying focus on new technological innovations. In the past, automation focused on the automation of manual labour (the First and Second Industrial Revolutions) and the automation of information processes through digitization (the Third Industrial Revolution). However, future automation processes will mainly focus on the automation of the work processes of the knowledge workers.

Today, Asia in general, and China in particular, have taken up the technological competition with the West. In China, the development and application of AI and robotics has become a decisive factor for increasing economic growth. This is evident in the increasing competition and tensions between China and the US (Allison, 2018). Against this background, the conflict we see developing between China and the US is in many ways focused on becoming the leading nation in the development of new technology in general, and AI and robotics in particular. The leaders of both countries are aware of the fact that the future leading nation, economically and militarily, will be the nation that takes the lead in the development of these technologies, specifically AI and robotics.

The future use and application of intelligent robots in the workplace will lead to a dramatic increase in the rate of change, which may be likened to a raging river after the spring snowmelt, devastating everything in its path and reshaping the landscape. We have witnessed similar upheavals in the three previous industrial revolutions. However, the new technology (AI and robotics) of the upcoming industrial revolution will not only change our workplaces, but the rate of change will be dramatically increased compared to previous industrial revolutions.

Economic growth

Automation not only leads to increased digitization and informatization of workplaces and increased productivity but also to stronger economic growth. Economic growth has been correlated with industrialization, since the period of the first industrialization from the 1750s onwards. One can say that industrialization is related to economic development and growth. This economic growth affects people in different ways. We demand more goods and services, we change our needs, and our expectations also change. However, economic growth not only affects people but also the planet. The unsustainable consumption of the Earth's resources is causing planetary crises, such as climate change and pollution. It is possible, as some critics claim, that economic growth is not the only factor causing a worsening climate change and pollution. However, since the First Industrial Revolution, air temperatures on Earth have been rising, contributing to climate change, although some commentators do not believe this to be a causal relationship. Economic growth is also closely linked to new skills and new knowledge. It might be said that the continual development of knowledge, during the periods of industrialization, has finally culminated in the development of artificial intelligence and the development of intelligent machines that will be able to create new knowledge. It can also be argued that it is precisely this new technology that can provide part of the solution to the current climate and environmental challenges that the world is facing. Paradoxically, we might propose that the economic growth that has created the challenges can also contribute to finding solutions to the challenges.

If this technology optimism, as expressed above, turns out to be correct, then the new technology will be able to promote economic growth and also provide remedies for the climate change and environment crises, as well as create a higher quality of life. We will return to how to facilitate such changes in Chapter 6, where we will develop various strategies to help people survive 'the robot shock'. Very briefly, however, we can suggest here that AI-powered systems can be used to increase efficiency and reduce carbon emissions. AI solutions can also be used to maintain species conservation and the planet's biodiversity and support various climate initiatives. In this way, artificial intelligence can create capabilities that will enable us to develop a greater degree of systemic competences, while at the same time contributing to sustainable economic growth. More specifically, this refers to Kate Raworth's doughnut economics (2018), which is based on the UN's sustainability goals (SDGs). In this context, AI applications can be used to achieve a new system of sustainable economic growth.

Economic growth has an explicit and implicit goal of developing new technology. There seems to be a clear connection between economic growth and new tools or between economic growth and technological innovations. When innovation has received such a considerable focus in our time, there are many indications that it is precisely because innovation, and especially technological innovations, that can lead to increasing economic growth. In the next section, we will examine the relationship between economic growth and economic inequality; this relationship has been thoroughly documented.[9] However, it is also a fact that economic growth has led to improving the living standards of more and more people. There is thus a clear relationship between economic growth and improvements in the standard of living and between economic growth and economic inequality. An important point in this context, however, is that economic growth has led to the vast majority of people improving their standard of living.

The new technology, that is, intelligent robots and artificial intelligence, will promote economic growth, but will also affect our workplaces. The future changes will have fundamental consequences for most people. Many commentators write about a coming wave of unemployment, which will result from the implementation of the new technology.[10] In this book, we argue that the consequences of robotization will be seen in the wage statistics, but not so much in the unemployment statistics. This is evident if we consider the low-paid Amazon workers (Delfanti, 2021). Other indications suggested by Graetz and Michaels (2018) suggest that when productivity increases, this can lead to the creation of new jobs. The statistics show no clear connection between robotization and mass unemployment.

Economic inequality

In recent decades, economic growth has promoted economic inequality (Allen, 2009). However, although economic growth results in greater economic inequality, economic growth also improves the standard of living of the poor. This insight is important to be aware of when addressing how to improve people's living standards during times of major changes. However, if economic equality and an improved standard of living is the goal, then economic growth is essential, because without economic growth there will be less for people to share. Although economic growth in the present day leads to greater economic inequality, this is not due to some unexplainable function of the economy, but due to the political structures that are geared towards increasing economic inequality. There can be many reasons why such a political economic ideology is implemented. In the book, *An Inquiry into the Nature and Causes of the Wealth of Nations*, first published in 1776, the political economist, Adam Smith, launched such a political economic ideology. Adam Smith is seen by some as being the 'Father of Modern Economics', as well as the 'Father of Capitalism'.[11] In other words, the field of modern economics is still greatly influenced by his ideas. The book developed the theory of free markets, division of labour and the importance of self-interest for economic growth. The point being made here is that it is not economic growth as such that leads to economic inequality, but the political and economic ideologies that govern economic and social development. If the goal is to reduce economic inequality, then it is the social mechanisms that promote economic equality that should be focused on, not less economic growth. It is in this context that Raworth's doughnut economy (2018) is of interest, because it shows which social mechanisms can be used. If one is not aware of the distinction between a phenomenon, such as economic growth and the political processes that lead to and result from this phenomenon, it will be difficult to address the relationship between economic growth and economic inequality.

It is the distinction between a phenomenon and the phenomenon's origin and consequences that is crucial to gain insight into, if one is to avoid confusion, frustration, fallacies and political disorientation. It is the same with robot shock as with economic growth. Robot shock is a phenomenon with causes that have led to the development of new technology and the consequences of this technology. Robot shock is a phenomenon that can lead to fewer jobs or increased productivity so that more people can find work and do what they are good at. Robot shock in itself neither eliminates nor creates new jobs. The political and economic ideology that is implemented in relation to the future development of robot technology will create guidelines for the social and economic consequences that will emerge. In other words, it is the information that is available and easy to access that will be utilized. This relates to the concept of 'availability cascades': This refers to the idea that we are controlled by the image of reality created by the media, because this image is easy to retrieve from memory. This may be expressed as follows: The more easily information enters into our consciousness, the greater the likelihood that we will have confidence in that information. In other words, we believe more in the type of information that is available in memory than the information that is not so readily available (availability proposition).

When we have separated the phenomenon from its causes and consequences, it will also be more straightforward to see how the consequences can materialize. As a result of the applications of artificial intelligence and intelligent robots in an economy, any possible negative developments will become clearer. It is, amongst other things, here, systemic thinking can come to our aid, which we elaborate on in Chapter 6.

The unregulated development of new technology or where technology is allowed to freely develop in a so-called 'free market', characterized by a global division of labour and global competence networks, where 'rational' self-interest is a dominant mechanism, will ultimately lead to increased economic inequalities and social challenges that will not benefit anybody. In this context, we can refer to Virginia Eubank's book (2017) *Automating Inequality*. She points out that unregulated high-tech development and applications often lead to greater economic inequality, where it is especially the poor who are 'punished' by agencies utilizing such technology indiscriminately.

When people start to feel inadequate or inferior in relation to mastering new technologies or tackling new cultural or political developments, this can lead to negative psychological and social consequences. We have seen this in the US when many jobs disappeared to China. Many people became alienated (Carney, 2020), disempowered (Frank, 2020) and politically disoriented (Frank, 2004). It seems reasonable to assume that something similar will occur when the consequences of robot shock are felt in people's workplaces. This may also lead to a situation with a breakdown in expectations regarding an education system that is failing young people, and where young people, instead of preparing for 'a good life', will end up being no better off than their parents (Ainley, 2014, 2016).

Robot shock is related to a time perspective, which we have witnessed in previous industrial revolutions. The development and implementation of AI and robotics in most aspects of the economy is not something that will occur overnight like some kind of technological phoenix bird. It will take a long time to develop the technology, and it will take a long time to implement the technology in most work processes. It will also take a long time before this technology becomes 'common property'. Economic growth and economic inequality are only one aspect of robot shock. Other aspects are connected to what Schumpeter (1951, 1954, 1989) terms 'creative destruction'. The 'creative destruction' caused by the robot shock will certainly lead to most work processes changing.

Changing work processes

Automation is here to stay, because increased productivity promotes economic growth. When an individual's work functions are subjected to more or less continuous automation processes, he/she will experience change as a constant factor. The introduction of new technology in the future will not only automate the work functions of blue-collar and white-collar workers but also, to an ever-increasing degree, the work functions of knowledge workers. In other words, workers that are involved in information, communication and knowledge production, as well as innovation workers. It is when these automation processes reach most people's workplaces that we can talk about 'changing work processes'.

Dumouchel and Damiano (2017) are most probably correct when they write that social robotics will become the new reality. In other words, social robots that can interact with humans on emotional and ethical levels (Johannessen, 2021a). It is also probable that we will see the creation of intelligent robots that can drive forward innovative ideas (Johannessen, 2022c). Regardless of how intelligent robots will be developed in the future, the constant search to increase productivity will drive forward the development of intelligent robots in order to promote the productivity of people's work processes. It is this search for ever more productive robots, which will result in automation becoming a permanent feature of everyday life. In other words, the development of intelligent robots that can increase productivity will be the driving force in the development of the new technology (Frey, 2019, 2020). In addition, AI and robotics will be increasingly applied in military operations.

Regardless of what will drive forward the development of intelligent robots and what direction this development will take, the fact remains that we will all have to deal with intelligent robots in our workplaces. Of course, it is not the case that these intelligent robots will resemble humans. It is only in certain work situations we may see the emergence of science-fiction-like humanoid robots. In most work processes, people will perhaps not even be aware of the fact that many of the work processes are controlled and operated by intelligent robots. These robots will most likely be nano-sized, such as nanorobots located in the networks people use. In other words, we will not 'see' the intelligent robots, but we will use them to increase the productivity and quality of the work processes we carry out.

It may be envisioned that intelligent robots will be used in student counselling to help students make choices about their future educational path, choose courses for them and decide how to put together entire educational packages. The intelligent robots will be able to do this, because they will 'know' the individual student's development history, expectations, opportunities and needs. The intelligent robots will also be able to predict, with greater accuracy than any human, the course of the student's future development. Against this background, it is highly probable that AI will be used in the future in student counselling services and other types of counselling. Although it will be a human who is formally in charge of the work processes, in reality, it will be intelligent robots, perhaps the size of a pinhead, that will drive the work processes forward. In fact, the use of AI and robotics in this way is not so much a future scenario, but something we can already find today in various sectors, such as the healthcare industry, the pharmacy industry, banking, the finance sector and in the military.

Such intelligent robots will be part of most work processes. It will seem that the individual worker doing his/her job is extremely productive, when in reality it will be an intelligent robot, or more likely intelligent robots interconnected in networks, that drive productivity forward.

If you ask people: 'What is a robot?', few will be able to give you a clear definition. However, one might say that a robot is a machine or device capable of carrying out complex work processes automatically in order to increase productivity. Another definition might focus on the use of algorithms. Algorithms are used in intelligent robots to facilitate planning and decision-making; in addition, intelligent robots also use learning algorithms, often interconnected in a larger network.

AI robots can be found in kitchens, helping people to make quality meals with greater precision and efficiency. We can also find these robots in the most sought-after surgeon's toolbox to enable surgeons to carry out robotic-assisted procedures. Robots can also be found in military aircraft and drones to improve military capabilities. For the most part, these intelligent robots are almost never visible to the naked eye. The only thing we can 'see', when intelligent robots are involved in work activities and processes, is the results that the robot has made possible.

Social and emotional robots will possibly be more visible than the intelligent robots that drive forward and change our work processes. However, it can be envisioned that social and emotional robots will be utilized in networks that will not be visible to us. For example, it may be imagined that such robots will be used in a service, where the purpose is to bring together people struggling with loneliness. It is quite possible that it will be precisely such networks that social and emotional robots will be able to perform their best services for people in the future.

The point of this brief introduction to changing work processes related to intelligent robots is to highlight the fact that these technologies will very likely operate in the background without them being visible to us. The intelligent robots can best be understood as a type of nano-technology. These will be located in our technological networks and promote our productivity, while continuously automating our work processes (Figure 2.1).

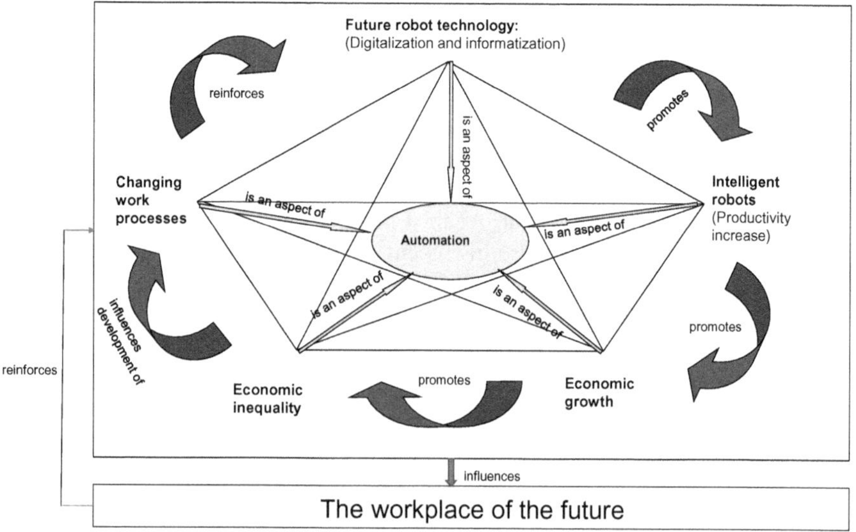

Figure 2.1 Automation and the workplace of the future: A conceptual model.

Analysis related to automation

This section will focus on automation related to robotics and AI rather than automation as a general phenomenon. However, by examining the history of automation in the first three industrial revolutions, this might reveal some general features regarding all types of automation processes.

One general feature is the opposition to automation processes. The most well-known example of opposition to automation is the Luddite movement.[12] Prior to the First Industrial Revolution, most weaving was done by domestic workers. However, the introduction of weaving machines resulted in the domestic weavers losing their jobs. This created a volatile situation. The weavers banded together under 'General Ludd' and became known as the 'Luddite weavers'. They carried out militant actions raiding the new textile factories and destroying the weaving machines. In other words, it is not difficult to understand the actions of the weavers as they lost their livelihoods. However, we might say that this is only one side of the Janus face of automation revealing the short-term perspective. The other side of the Janus face, the long-term perspective, shows that eventually more new jobs were created, productivity increased and there was a general rise in wages. The new workplaces were generally safer and created opportunities for more and more areas of expertise than before. The objection to this view is that it took time before the new workplaces became a reality, so it is doubtful that this long-term perspective benefitted the Luddite workers who lost their jobs. In other words, there is a time lag between the short-term and the long-term perspectives, regarding the effects of automation processes. Thus, it took time from when the jobs of the Luddites were destroyed, until new jobs were created in new sectors of the economy. In the future, it will be important to introduce strategies to counteract the negative aspects of this time lag in connection with the 'robot shock'. It is this that we will focus on in Chapter 6.

Resistance to change thus appears to be something we see in all types of automation processes in the economy. It therefore also seems reasonable to assume that as we move towards the Fourth Industrial Revolution, where AI will constitute the core technology, automation processes driven by new AI applications, will trigger resistance to changes. Consequently, it will be important to develop competences and cultures in businesses and organizations, in order to strategically tackle this future opposition to changes. The question that emerges, however, is the following: Can there be something in the idea that the new technology will negatively impact people's workplaces? The question is straightforward enough, but the answer is by no means clear. First, we need to distinguish between the short-term and long-term perspectives. In a short-term perspective, it is highly likely that many jobs will disappear, change and be transformed into something new. Some functions will be lost or taken over by intelligent robots. It is conceivable that Daniel Susskind (2020) has a point when he says that a situation will emerge in the future where there won't be work for everyone. However, we can also say that this only represents the short-term perspective. In a long-term perspective, the three previous industrial revolutions have shown that more jobs were created than those lost. Moreover, between the Second and Third Industrial Revolutions, during the second half of the 20th century, public secondary education was introduced, enabling young people to gain new skills that were adapted to new work functions and new jobs. In other words, in the longer term, economic history shows that technology ultimately creates more new jobs than those that are initially automated away. However, in this context, one might cite John Maynard Keynes' when he ironically noted, 'In the long run we are all dead',[13] that is, the economists who talk about the 'longer perspective' and about how things will eventually perk up. This may be all very well, but what about the worker who loses his/her job to automation? Should he/she just wait around for this new bright future to emerge? In other words, the academic, the politician and the policy developer, may be interested in distinguishing between the short-term and long-term perspectives, but is this of interest to the person who loses his/her job, and who is negatively affected by the consequences of automation? The title of a book by Geoff Mann, *In the Long Run we are all Dead: Keynesianism, Political Economy and Revolution*, (2019) implies the point that was also made by Keynes that economists and policymakers need to take the different perspectives into consideration in relation to time lag. Mann focused his analysis on the 2007–2008 financial crisis and its aftermath and which policies were introduced to deal with the crisis in relation to different perspectives. However, this way of thinking can also be applied regarding the impact of AI and robotics on the automation of work functions in the future and how to deal with any negative short-term consequences.

Keynes and Mann both importantly point out that the short- and long-term perspectives should be integrated through different system levels. This is not to save or write off capitalism as such, but to help people and social systems that are in crisis. The crises occur because people and organizations have to relate to the short-term perspective more than the long-term perspective. But dealing with the short-term negative consequences is often something that is beyond the sole capacity of people and organizations. Consequently, the state needs to intervene and provide assistance in situations that arise, both in the short-term and long-term.

Mann (2019) astutely points out that when the left-wing of the political spectrum supports Keynes-like remedies to capitalist crises, they have, in reality, stopped believing in an alternative to the capitalist system. Mann mentions that the government interventions, during the 2007–2008 financial crisis, in order to keep troubled banks afloat, in reality, also kept capitalism as a system afloat. Karl Marx believed that capitalism would eventually destroy itself due to internal contradictions (1976, 1978, 1991). This may turn out to be correct. However, as mentioned above, the capitalist system has come to rely on government interventions in order to deal with its internal contradictions. In relation to Mann's (2019) view, governments will probably intervene to tackle the probable future crisis that will result from the robotization and automation of workplaces. The paradox here is that such government interventions will most probably be supported by liberals and social democrats, in order to prop up the capitalist system.

Keynesianism, in Mann's view, which agrees with Dorling (2019), is an economic strategy that maintains the 1 per cent society. The 1 per cent society is a society where a minute fraction of the population –l ess than 1 per cent – controls the lion's share of the wealth, power and resources, and where economic inequalities are extremely large. This may be compared to the hierarchical structures that were evident during the feudal society. Consequently, we find it appropriate to term this new emerging 1 per cent society feudal capitalism (Johannessen, 2023).

When economists and politicians join forces to save this feudal structure by intervening with Keynesian management tools, they also help to maintain this social inequality (Picket, 2017). In this context, it is interesting to envision how intelligent robots and artificial intelligence, and automation resulting from this technology, will affect the development of future workplaces. One plausible scenario, which follows from Mann's and Dorling's descriptions, is that a new elite will emerge that will further maintain the 1 per cent society. This elite will probably consist of those people with extensive high-tech qualifications that are compatible with the new technology. These qualifications may be linked to STEM competences.[14] Most likely, this competence elite will also include people with social science qualifications. It is also probable that these people will support and argue for the maintenance of a capitalist system that reinforces and preserves the 1 per cent society, the new feudal capitalism. Specifically, this may also include professions such as economists and lawyers, as well as managers at various levels. If such a scenario is realized, a new form of automation will emerge, where inequalities will also become automated (Eubanks, 2017), and feudal structures will be re-established (Kotkin, 2020; Johannessen, 2023). What we can see developing are automation processes becoming established and new elites becoming established at the top of the social ladder. The new technology is thereby contributing to freezing fast social structures, so that automation not only maintains inequality, but reinforces it and lays the foundation for a completely new capitalist system, feudal capitalism (Johannessen, 2023). The workplaces of the future will be structured by feudal capitalism, the 1 per cent society (Dorling, 2019). This will result in closed divisions between those who govern and those who are governed. In addition, feudal capitalism, a form of representative democracy, will utilize the new technology to create very specific perceptions of reality amongst the population. By using the new technology to communicate, manage, control and coordinate activities and processes, one also manages to create the understanding of reality that is most useful, seen from the perspective of the 1 per cent of the population at the top of the social hierarchy. It is in this situation that we can say that we are all controlled by the image of reality created by the media, because this image is easy to retrieve from memory. This concept is termed 'availability cascades'. This may be expressed as follows: The more easily information enters our consciousness, the greater the likelihood that we will have confidence in that information. In other words, we believe more in the type of information that is available in memory, than the information that is not so readily available. This is referred to as 'the availability proposition'.

Theoretical reflections related to automation

It seems reasonable to assume that the automation of workplaces in the future will be dependent on two factors. The first is the digitization of work processes. This digitization manifests itself in what we have termed informatization. Informatization refers to all the processes in an organization that are based on data, information and knowledge processes. In order to illustrate what is meant by informatization, it may be advantageous to contrast it with the automation of production processes. In a traditional industrial operation of the past, the production processes were separated from the administrative processes to a large degree. It was chiefly the material production processes that were automated in the first two industrial revolutions. The Third Industrial Revolution involved the transition from analogue to digital technology. In other words, IT technologies and information processes became more involved in the automation of industries, businesses and organizations. In the Fourth Industrial Revolution, it will largely be the information processes that will be automated.

In the past, it was first the blue-collar jobs that were automated using various mechanical and power technologies. Later, white-collar jobs were also automated using digital technology. The automation of jobs in the future, in the Fourth Industrial Revolution, will mainly use AI and robotics technologies (Schwab, 2016, 2018). This will constitute the second of the two factors regarding the automation of workplaces in the future. This will mainly concern the automation of white-collar jobs and more specifically the jobs of knowledge workers. It is in this context that information processes will constitute an important part of the production system.

In addition to the two factors mentioned above, i.e., informatization and AI and robotics, new skills and expertise will be needed for developing, applying and making use of the new technologies. In other words, the development of the necessary skills and expertise will constitute the very foundation regarding the automation of future workplaces.

It cannot be stressed often enough that the First, Second and Third Industrial Revolutions did not occur 'overnight'. Susskind rightly points out the revolutions were 'not a big bang, but a gradual withering' (2020: 3). The time perspective is important here. Although the future Fourth Industrial Revolution is termed a 'revolution', the word 'revolution' is a misnomer and gives a misleading impression regarding the nature of future economic change. It would be more correct to use the word 'evolution'. In other words, the changes will not occur with a 'big bang' but will involve a gradual development. In practice, this means that most people, organizations and societies will have time to adapt to the automation of future workplaces. However, it is crucial that people learn to adapt to the economic reality that will gradually emerge. The objection to this view is of course that no one knows exactly what will emerge in the future. But from economic history, we do know some of the central features that have created social processes in the past. We know that new technologies have always influenced and created the conditions regarding new work processes. We have seen this since the invention of the wheel, the invention of the compass, the invention of the Spinning Jenny, and more recently, with the development of the Internet. There is nothing to suggest that the new technology that is emerging will not affect the workplaces of the future. However, no one knows exactly what the final outcome will be. What we do know, however, is that new skills will be in great demand.

In the past, it has always been the skills and expertise that are compatible with the emerging technology that have emerged victorious from the competition. In the Fourth Industrial Revolution, there may be many technologies competing to dominate various processes in working life. However, it is intelligent robots and digitization that we are focusing on here. In this connection, our main proposition is the following: The greater the degree of intelligent robots that are used in work processes, and the greater the degree of digitization that is used in these work processes, the greater the probability that these workplaces will be automated.

The proposition is based on economic historical insights from the preceding industrial revolutions. In these revolutions, the new technology may have been used in many work processes, but in all these work processes automation was paramount (Frey, 2019: 141–189).

Automation can be implemented at different levels and in relation to different work processes. Productivity growth is often the most relevant factor. However, we should not ignore the fact that automation, from a historical perspective, has replaced much of the drudgery and dangers associated with manual labour. When these work processes have been replaced, the workers will of course also have lost their jobs. In a short-term perspective, this may be correct. In a medium-term and long-term perspective, new jobs will have been created, demanding new skills and expertise. For instance, taking a long historical perspective, there are millions more today doing paid work than at the time of the First Industrial Revolution (1750–1840). In other words, we cannot escape the fact that automation, new technology and new skills create new jobs. Joseph Schumpeter, when writing about innovation theory, uses the concept of 'creative-destruction' (2002, 2006, 2010). Creative destruction refers to continuous and disruptive innovation processes, leading to new production methods replacing outdated ones. In this context, the 'old' jobs are 'destroyed' by new technology and automation processes, while new jobs are created. Viewed in the long term, this development leads to a better life for more and more people.

Technology that promotes productivity growth and automates workplaces is, as shown above, nothing new. It is possibly the most important aspect of the new technology in the workplace that it is leading to fewer and fewer labour-intensive jobs. New AI and robotics applications are resulting in the automation of more work functions than before. In this context, many critics have discussed the possibility of this resulting in mass unemployment in the future (Susskind, 2018, 2020; Ford, 2016, 2021). Of course, this may very well occur. On the other hand, the future innovation economy, driven by new innovations and new technology, will result in a demand for new skills and expertise, and especially for people with creative skills.

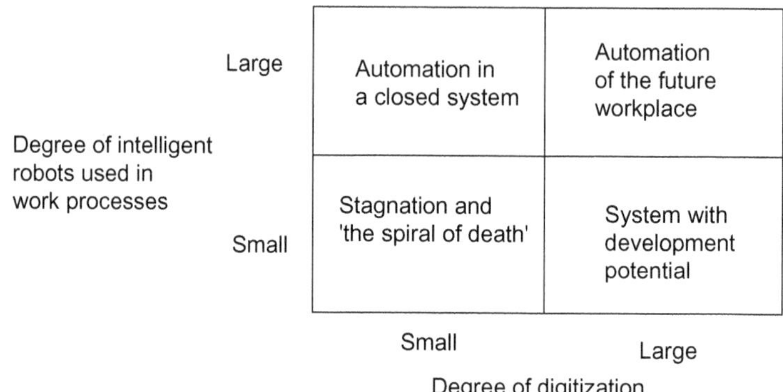

Figure 2.2 Automation of the workplace of the future: A typology.

There are many factors that can lead to unemployment, such as long-term inflation and economic depression. On the other hand, new technology, new organizational innovations and other types of innovations lead to productivity improvements and economic growth (Krugman, 1995: 56). New skills and expertise that are spread to the various sectors of the economy are not a danger to employment, quite the opposite. New jobs, new work functions, new expertise and skills have, throughout economic history, gone hand in hand with the increased automation of workplaces. Based on this thinking, new automation processes are not a danger to employment, on the contrary. New automation processes result in changes in the workplace, which may be experienced as frustrating for some people and result in others losing their jobs. However, throughout history, new work functions and new jobs have also followed in the wake of new technology and automation of workplaces.

Based on the above narrative, description, analysis and theoretical reflections, we have developed a typology for the automation of the workplace of the future (Figure 2.2).

Practical utility related to automation

It has largely been routine jobs, dangerous jobs and jobs where health, safety and environment (HSE) challenges have been involved, which have previously undergone automation processes. However, with the new developments in AI and robotics, completely different jobs than just the routine ones have been exposed to automation. Digitization and informatization have led to many jobs related to information and communication processes becoming exposed to automation. For instance, AI has been utilized in the banking and finance sectors to automate market analyses and for making decisions about buying and selling shares. All the activities and processes that can use algorithms can also be automated by AI applications. It is no longer necessary to program machines, so that they can perform simple and complicated procedures. Intelligent robots have now reached the point where they can learn from their own mistakes. The more we expose these robots or learning algorithms to events in real situations, the more they learn, and the greater the probability that these robots can perform the operations better than humans. This may sound like science fiction, but it is already happening. For instance, in the game of chess, where computers can learn from playing chess against themselves. In every imaginable job function, learning algorithms can develop 'skills' through testing out solutions in safe environments that enable them to outperform the best human experts. This applies to sectors and fields such as health, research, education, creative processes, idea development and innovation. When, not if, learning algorithms are implemented in various intelligent robots that focus on specific tasks, automation will become an important part of all the work functions in working life. It is possible that senior management jobs and jobs that involve strategic decision-making will not be affected to the same extent. In these jobs, it may be conceivable AI and intelligent learning algorithms will be used as support functions. However, in all other areas of work, it is conceivable that automation will become an important part of work processes.

The automation of work functions can be related to two factors.

First, when new technology directly replaces people's work functions, these people will have to find other work functions or jobs. In addition, they will most probably have to learn new skills.

Second, the application of new technology may not replace work activities but create new activities related to the established ones. For instance, new applications that can enable researchers to perform their jobs better and faster. New technological applications will enable trades workers, such as plumbers, electricians and carpenters, to perform their work better and at a lower cost. In these examples, the new technology will not replace workers but enable them to do their jobs better.

In other words, new technology will both replace and enable the improvement of existing work functions. In this way, we can say that automation is not so much a linear function that will replace the human workforce. Against this background, it is not certain that the automation of future workplaces will lead to mass unemployment. As mentioned above, some workers will have to learn new skills and find new work functions or jobs, while other workers will be able to utilize the new technology to augment their skills, so they can be more productive. Thus, it may be said that productivity will be the key factor regarding the introduction of new technology. The technological applications that replace people's work functions will increase the system's productivity. The work processes that use the new technology to improve the execution and quality of established work processes will also increase productivity. One can therefore say that the new technology has the potential to increase both work productivity and system productivity.

The future social costs of productivity growth will mainly be related to human labour that is replaced by machines, i.e., AI and robotics applications. However, these social costs are transitory, because productivity growth caused by automation, will, after a certain time lag, create more jobs than it replaces. The key word here is time lag. In the period between when new technology is implemented, to when new jobs are created, many people will be negatively impacted by productivity improvements. In societies that have not developed welfare safety nets for such transformations, many people will not only lose their jobs but also their income. On the other hand, societies that have developed effective welfare safety nets will be able to replace some of the lost income, so that families will be able to fill their basic needs. This has two positive sides. First, the loss of a job will not be the same as the loss of all income. Second, their income from welfare payments will help maintain demand. Without this income, the demand for goods and services would drop significantly. Drop in demand could negatively affect other sectors leading to more unemployment. In such a situation, the negative consequences of productivity improvements could affect far more people than those who lost their jobs due to new AI and robotics technology.

Regardless of how societies have protected, or not protected, their citizens from the possible negative consequences of increased productivity, organizational innovations will also emerge. The rationale is that systems do not function effectively when they are exposed to qualitatively new technologies, while still using organizational forms that are more suited to the old technology (Antonelli, 2009; Frey, 2019).

We envision a new way of organizing developing in several steps. The first step could be a greater focus on what the system is designed to do. This question was investigated by Stafford Beer (1995). Beer's purpose was to organize social systems in relation to what they are designed to do. In an organization's development, many functions will emerge that are not really related to what the system is designed to do. The history of organizations has shown that as a system grows and is able to cope with the increased cost of developing more functions unrelated to the system's primary task, many such functions will continue to emerge (Gratton, 2009, 2012). When new technology promotes productivity, focus will also be directed towards the functions that are not really related to what the system was designed for. The rationale is not related to costs, because the income of the organization is likely to grow as a result of increased automation. The reason is that they now have opportunities to focus on organizational innovations. This is mainly explained by the fact that if competitors introduce organizational innovations, then the relative profit will come under pressure. Thus, they may risk losing in an open market, where the search for the best profit opportunities will control much of the investment capital flowing to the company in question. In other words, new opportunities will emerge for organizational innovations. It seems reasonable to assume that these will be structured in relation to the front line – where the goods and services are produced and delivered. This means that the expertise in the front line will be given information, decision-making authority, resources and opportunities for increased efficiency and earnings. Another consequence of the new organizational innovations will probably be a greater focus on doing what the system was designed for. Thus, several of the functions that do not fall under this category will be de-prioritized, and in some cases discontinued. In this way, we will get more flexible organizations that can adapt more quickly to changes in the environment. With this reflection, we can say that businesses will automate many of their work processes and discard functions not related to their primary task. Automation of

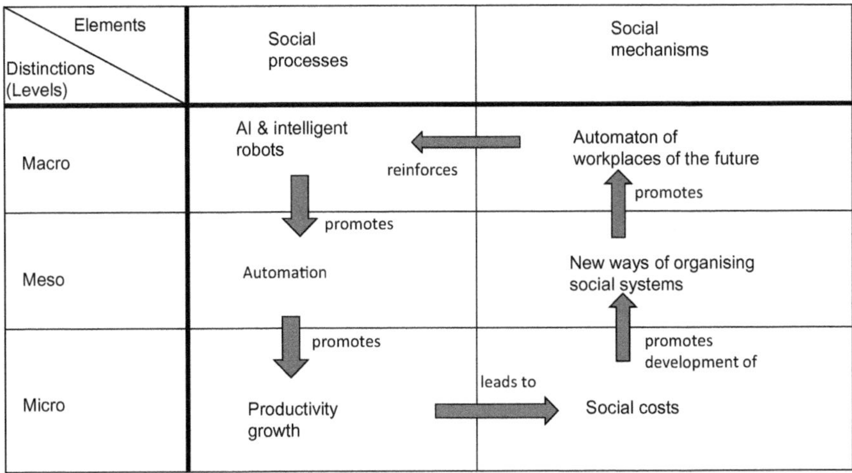

Figure 2.3 Automation of future workplaces: A Boudon-Coleman diagram.

the workplace of the future will, with such a reorganization, also lay the foundation for a reinforced development of artificial intelligence and intelligent robots in work activities and processes.

Based on the above narrative, description, analysis, theoretical reflections and the review of the utility value, we have developed a Boudon-Coleman diagram for the automation of future workplaces (Figure 2.3).

Systemic connections related to automation

New knowledge is gradually discovered until eventually a certain threshold is reached, breaking down boundaries, and resulting in a technological storm, or what we may term a technological revolution. Of course, no one knows when this threshold will be exceeded. What is known, however, is that when the threshold is exceeded, it is as if a dam breaks, resulting in a flood of new ideas.

In this context, the development of artificial intelligence can be viewed as occurring through various processes. For instance, these processes can be related to the development of data storage, various types of algorithms, databases, new forms of management and control, and so on. The sum of these processes has resulted in new innovative ideas that eventually culminated in what we today term artificial intelligence. However, before this point was reached there were many problems that had to be solved. For example, 'cloud storage' enabled users to store data and files on the internet, so they had access through the internet, and were not dependent on storing data at a physical location. Thus, millions of users created a demand for storage and processes in the network itself, i.e., not at a physical location. Gradually, increased demand and an increase in technological services also created the need for AI applications in networks. In relation to these developments, there was also an increased digitization of work processes, such as the conversion of analogue data into digital data. When digitization of work processes became a reality, this created a demand for new skills in order to develop, operate, maintain and secure digitization processes. Thus, the automation of work that digitization resulted in also created a demand for new skills. In this way, qualitatively new jobs were created, while old working methods were destroyed. In other words, new technology destroys old working methods, while at the same time creating new ones.

Bottlenecks also develop in this automation process. In the example of digitization, some regions are less developed than other regions. This restricts the opportunities of some regions to adopt the new technology as effectively as those regions where the conditions are better. Specifically, this could concern the development of fibre networks. Those regions without effective fibre networks will not be able to exploit the technological advantages that others have access to. Bottlenecks are directly linked to time lag. It will take some time before some regions have the same opportunities as other regions. This time lag may also be related to a 'competence gap', because those regions that are more developed will attract people with high-tech skills. In this way, some regions can end up as technological backwaters due to lack of technological development. This process is reinforced when new expertise is attracted to those regions that have the best access to networking, business development and so on. Therefore, such bottlenecks can pose a serious problem for technological development and can very easily lead to some regions being left behind.

Both automation and bottlenecks may reinforce the development that some people will become winners, while others will become losers, when the new technology impacts the future labour market. This can lead to negative social developments. In the case of 'robot shock', the negative social costs can be related to a collective blindness where those who are responsible for social and economic development wash their hands of the matter and leave new developments up to the market. The point here is that the market does not solve problems, people do and the upcoming 'robot shock' will be one such crisis for those people who are exposed to its consequences. The crisis may be felt in many ways. One way in which this crisis can emerge is that many bottlenecks can occur together. First, a bottleneck may be related to lack of technological development; second, a bottleneck can be related to lack of competence; third, another bottleneck may be related to the lack of development of new businesses based on the new technology; fourth, yet another bottleneck may be related to a lack of innovation; fifth, a bottleneck may be related to poor wage growth, and so on. Numerous such bottlenecks may result in some areas and regions lacking the impetus to develop business growth. In this situation, there will thus be a danger that these areas will become long-term technological and economic backwaters. In these areas and regions, it can be said that automation will have a negative impact on the development of future workplaces. A proposition for a social law in this context is the following: The areas and regions that end up in the backwater of technological and economic development will very likely remain in this backwater.

Bottlenecks can occur in many ways when new technology is developed. One of these bottlenecks can concern the demand for new skills. In practice, this may concern the operators of intelligent robots, algorithm developers, computer programmers, etc. When businesses in one region of a country want to adopt the new technology, in order to promote automation and productivity, they may encounter problems with a bottleneck related to lack of competences. In such a context, these businesses either do not adopt the new technology or that they move their operations to locations where it is easier to implement the new technology and find workers with the appropriate skills. If such bottlenecks are not resolved, they can damage business development and the job market in the regions where they are allowed to become part of the region's business structure. In such cases, it does not seem probable that the market can remedy such bottlenecks; it is rather the public authorities that need to deal with them. Therefore, it can be said that the public authorities need to regulate the market, so that everyone will have the opportunity to take part in the development of the new technology.

Throughout history, there have been economic growth bottlenecks due to skills shortages (Frey, 2019: 240). It is precisely in this context that governing authorities can play a decisive role. Although it is difficult to know exactly which skills will be in demand in the workplaces of the future, we know that they will need to be compatible with the new technology (Aghion et al., 2021; Susskind, 2018). Of course, we do not know exactly which new technologies will become dominant in the future, but it is very probable they will be related to AI and robotics (Harari, 2018). Thus, skills related to AI and robotics, such as skills in coding, algorithms and digitization, will most probably be in demand. It is also probable that bottlenecks will develop due to a shortage in these skills, resulting in an increased demand (Susskind, 2020).

The bottlenecks related to skills shortages will create winners in the future labour market. The winners will be those people with skills compatible with the new technology (Johannessen, 2024). The losers will be those who do not develop new skills, but rely on their 'old' skills (Johannessen, 2025). This will result in negative social costs related to low wages, amongst other things. The negative social costs in a wider perspective will be related to people's breakdown of expectations regarding future opportunities for themselves and their families. Research in the US and England has shown that intelligent robots reduce the opportunities for unskilled people to find a job (Frey, 2019: 342–243). In practice, this means that those people without a college education will not have the same opportunity in the job market as their parents had. Those who have a college education that is not compatible with the new technology will also experience problems (Ainley, 2014, 2016).

Like in the past, education will also be important in the future in order to keep up with the demands of new technology in relation to skills in the labour market (Acemoglu & Autor, 2011). However, the new intelligent robots are taking over so many functions in working life that the demand for skilled and specialized labour will be limited (Ainley, 2016). If this assumption is correct, people with higher education levels will no longer be guaranteed a secure and well-paying job. In other words, there appears to be emerging an uneven race between the development of technology and the development of education that meets the demands of the new technology (Goldin & Katz, 2008). Thus, education is struggling to keep up with technology. If the individual feels that it is not possible to win this 'race', then the question arises why he/she should invest money, time and commitment on something that will not produce results (Ainley, 2016). For the individual, not taking part in the 'race' will also mean that he/she will only be able to get an insecure and poorly paid job in the future labour market. For society, this situation will be tantamount to losing talent and not keeping up with the skills' revolution, which is directly linked to the development of the workplace of the future. Losing the competition for acquiring the skills of the future related to the new AI and robotics technologies, amongst others, will negatively impact, people, organizations and nations. It is in this context, one can understand the mounting tensions between the US and China, regarding the battle to be at the forefront of expertise in the new technologies (Allison, 2018).

The negative social costs of not participating in the race to acquire the new skills related to artificial intelligence and intelligent robots can result in consequences on many levels and for different parts of society. It can be said that this race shows the systemic connections between education and technology, which can have consequences not only for the individual but for businesses and society. The 'Sputnik shock' occurred in 1957 when the Soviet Union launched the world's first satellite, Sputnik I into the orbit around the Earth. This was a 'shock', because it was believed at the time that the US led the way regarding space technology. The 'shock' resulted in Congress vastly increasing education funding. Similarly, if

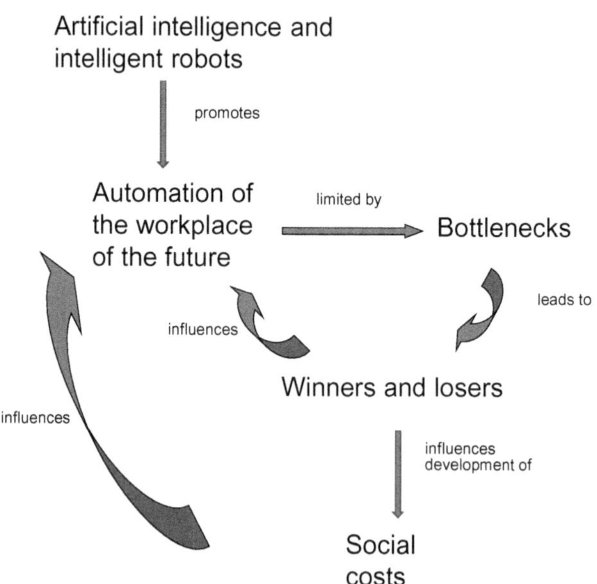

Figure 2.4 Automation and development of the workplace of the future: A loop diagram.

the US loses the race with China to develop new advanced technologies, this may result in a new 'shock', which will constitute a 'robot shock' for the American education system of the future.

Based on the above narrative, description, analysis, theoretical reflections and the review of the practical utility, as well as a review of the systemic connections, we have developed a loop diagram that shows how automation will affect the development of the workplace of the future (Figure 2.4).

Sub-conclusion related to automation

We have examined the following question: What is the potential impact of automation on future employment trends?

The short answer is that there will be a race to secure competence that is compatible with new technologies. This will lead to fierce competition for places at top universities. Inevitably, many people will be unsuccessful. As a result, we will see a future labour market in which graduates of top universities can secure well-paid jobs, while those who lost out will have to be content with insecure, lower-paid work. One consequence is that socio-economic inequalities will become even more extreme than they are today. At a national level, this race to secure competence and develop technology will lead to increasing competition, and possible conflict, between the world's two largest economies, China and the US.

Notes

1 https://www.globaltimes.cn/page/202111/1238203.shtml
2 Antonelli, 2009; Atkinson & Ezell, 2014; Billington, 2020; Brynjolfsson & McAfee, 2011, 2014; Brynjolfsson & Saunders, 2013; Chesbrough, 2003, 2006;Chesbrough et al., 2008; Hamel, 2002, 2007; 2008, 2012; Davenport, 2005, 2019; Davenport & Kirby, 2016; Drucker, 1999, 1999a, 2005, 2007; Johannessen, 2023; Kahan, 2013; Rogers, 2003; Skarzynski & Gibson, 2008; Tucker, 2002; Unterberg, 2013; Von Hippel, 2005; Zaltman et al., 1973.
3 Johannessen, 2020c, 2020d, 2021b, 2022a, 2022d.
4 https://en.wikipedia.org/wiki/Cybernetics
5 There are also technologies other than artificial intelligence that will drive the Fourth Industrial Revolution forward, e.g., gene-editing, gene-engineering, Big Data, block chain technology, etc. However, artificial intelligence will constitute the core technology of the Fourth Industrial Revolution.
6 https://en.wikipedia.org/wiki/Automation
7 Johannessen, 2020, 2020a, 2020b, 2021a, 2022a, 2022b, 2022c.
8 https://en.wikipedia.org/wiki/Luddite
9 Acemoglu et al., 2015; Atkinson, 2015; Dorling, 2015; Eubanks, 2017; Milanovic, 2016; Picket, 2017; Piketty, 2014, 2016; Stilwell, 2019.
10 Oppenheimer, 2019; Baldwin, 2019; Ford, 2016, 2021.
11 https://en.wikipedia.org/wiki/Adam_Smith
12 https://en.wikipedia.org/wiki/Luddite

13 Keynes stated this in an article in 1923. However, the entire statement was: 'The long run is a misleading guide to current affairs. In the long run we are all dead.' (https://finfacts.ie/Irish_finance_news/articleDetail.php?In-the-long-run-we-are-all-dead---John-Maynard-Keynes-159)
14 STEM stands for Science, Technology, Engineering, Mathematics.

References

Acemoglu, D. & Autor, D.H. (2011). Skills, tasks and technologies: Implications for employment and earnings, in Card, D. & Ashenfelter, O. (eds.), *Handbook of labor economics*, vol. 4, Elsevier, Amsterdam, pp. 1043–1171.

Acemoglu, D.; Suresh, N.; Pascual, R. & Robinson, J.A. (2015). Democratic redistribution, and inequality (Chapter 21), in Atkinson, A.B. & Bourguignon, F. (eds.), *Handbook of income distribution*, vol. 2, Elsevier, London. pp. 1885–1966

Aghion, P.; Antonin, C. & Bunel, S. (2021). *The power of creative destruction*, Harvard University Press, Cambridge, MA.

Ainley, P. (2014). Follow your dreams and attend to universities if possible, *Latitude*, 21 December.

Ainley, P. (2016). *Betraying a generation: How education is failing Young people*, Policy Press, Bristol.

Allen, R.C. (2009). *The British industrial revolution in global perspective*, Cambridge University Press, Cambridge.

Allison, G. (2018). *Destined for war: Can America and China escape Thucydides trap?* Scribe, London.

Antonelli, C. (2009). The economics of innovation: From the classical legacies to the economics of complexity, *Economics of Innovation and New Technology*, 18 (7): 611–646, doi: 10.1080/10438590802564543. To link to this article: http://dx.doi.org/10.1080/10438590802564543.

Atkinson, A.B. (2015). *Inequality: What can be done*, Harvard University Press, Cambridge.

Atkinson, R.D. & Ezell, S. J. (2014). *Innovation economics: The race for global advantage*, Yale University Press, New York.

Autor, D. (2015). Why are there still so many jobs? The history and future of workplace automation, *Journal of Economic Perspectives*, 29 (3): 3–30.

Baldwin, R. (2019). *The globotics upheaval: Globalization, robotics and the future of work*, Oxford University Press, Oxford.

Beer, S. (1995). *Diagnosing the system for organizations*, John Wiley & Sons, London.

Bennett, S. (1979). *A history of control engineering 1800–1930*, Peter Peregrinus Ltd., London.

Billington, D.P. (2020). *From insight to innovation*, MIT Press, Boston.

Brynjolfsson, E. & McAfee, A. (2011). *Race against the machine*, Digital Frontier Press, New York.

Brynjolfsson, E. & McAfee, A. (2014). *The second machine age*, W.W. Norton & Company, New York.

Brynjolfsson, E. & Saunders, A. (2013). *Wired for innovation: How information technology is reshaping the economy*, The MIT Press, London.

Carney, T.P. (2020). *Alienated America*, Harper, New York.

Chesbrough, H.W. (2003). *Open innovation: The new imperative for creating and profiting from technology*, Harvard Business School Press, Boston.

Chesbrough, H.W. (2006). *Open business models*, Harvard Business School Press, Boston.

Chesbrough, H.; van Havebeke, W. & West, J. (2008). *Open innovation: Researching a new paradigm*, Oxford University Press, Oxford.

Davenport, T.H. (2005). *Thinking for a living, how to get better performance and results from knowledge workers*, Harvard Business School Press, Boston.

Davenport, T.H. (2019). *The AI advantage*, The MIT Press, Cambridge, MA.
Davenport, T.H. & Kirby, J. (2016). *Only human need supply*, Harper Business, New York.
Delfanti, A. (2021). *The warehouse: Workers and robots at Amazon*, Pluto Press, New York.
Dorling, D. (2015). *Injustice: Why social inequality still persists*, Policy Press, London.
Dorling, D. (2019). *Inequality and the 1%,* Verso, London.
Drucker, P.F. (1999). Knowledge worker productivity: The biggest challenge, *California Management Review*, 41 (2): 79–94.
Drucker. P.F. (1999a). *Management challenges for the 21st century,* Harper Collins, New York.
Drucker, P.F. (2005). Managing oneself, *Harvard Business Review*, January, pp. 100–109.
Drucker, P.F. (2007). *Innovation and entrepreneurship*, Elsevier, London.
Dumouchel, P. & Damiano, L. (2017). *Living with robots*, Harvard University Press, Boston, MA.
Eubanks, V. (2017). *Automating inequality*, St. Martin's Press, New York.
Fang, N. & Fu, Y. (2020). *China's political system*, Springer, London.
Ford, M. (2016). *The rise of the robotics, Technology and the threat of mass unemployment*, Oneworld, New York.
Ford, M. (2021). *Rule of the robots*, Basic Book, London.
Frank, T. (2004). *What's the matter with Kansas? How conservatives won the Heart of America*, Metropolitan Press, New York.
Frank, T. (2020). *People without power*, Scribe, London.
Frey, C.B. (2019). *The technology trap: Capital, labor, and power in the age of automation*, Princeton University Press, Princeton.
Frey, C.B. (2020). Attitudes towards technology: Part II, in Skidelsky & Craig, *Work in the future: The automation revolution*, Palgrave, London, pp. 89–97.
Goldin, C. & Katz, L.F. (2008). *The race between education and technology*, The Belknap Press, New York.
Graetz, G. & Michaels, G. (2018). Robots at work, *Review of Economics and Statistics*, 100 (5): 753–768.
Gratton, L. (2009). *Glow*, Financial Times/Prentice Hall, New York.
Gratton, L. (2012). *Hot spots: Hot spots: Why some teams, workplaces, and organization buzz with energy-and others don'ts*, Read How You Want, London.
Guarnieri, M. (2010). The roots of automation before mechatronics. *IEEE Industrial Electronics Magazine*, 4 (2): 42–43. doi:10.1109/MIE.2010.936772. hdl:11577/2424833. S2CID 24885437.
Hamel, G. (2002). *Leading the revolution: How to thrive in turbulent times by making innovation a way of life*, Harvard Business School Press, Boston.
Hamel, G. (2007). *The future of management*, Harvard Business School Press, Boston.
Hamel, G. (2008). Introduction, in Skarzynski, P. & Gibson, R. (eds.), *Innovation to the core*, Harvard Business Press, Boston, pp. xvii–xix.
Hamel, G. (2012). *What matters now: How to win in a world of relentless change ferocious competition, and unstoppable innovation*, John Wiley & Sons, New York.
Harari, Y.N. (2018). *21 lessons for the 21*st *century*, Jonathan Cape, London.
Johannessen, J-A. (2020). *The workplace of the future*, Routledge, London.
Johannessen, J-A. (2020a). *Automation, innovation and economic crises: Survival and the fourth industrial revolution*, Routledge, London.
Johannessen, J-A. (2020b). *Artificial intelligence, automation and the future of competence at work*, Routledge, London.
Johannessen, J-A. (2020c). *Knowledge management for leadership and communication: AI, innovation and the digital economy*, Emerald, London.
Johannessen, J-A. (2020d). *Knowledge management philosophy: Communication as a strategic asset in knowledge management*, Emerald, London.

Johannessen, J-A. (2021). *China's innovation economy: Artificial Intelligence and the new silk road*, Routledge, London.
Johannessen, J-A. (2021a). *Artificial intelligence, automation and ethics in the innovation economy*, Routledge, London.
Johannessen, J-A. (2021b). *Communication as social theory: The social side of knowledge management*, Emerald, London.
Johannessen, J-A. (2022). *The new silk road and the innovation economy in China*, Routledge, London.
Johannessen, J-A. (2022a). *Creativity, innovation and the fourth industrial revolution: The da Vinci strategy*, Routledge, London.
Johannessen, J-A. (2022b). *A systemic approach to continuous change in the innovation economy*, Routledge, London.
Johannessen,J-A. (2022c). *Intelligent robots consciousness and creativity: The search for hidden knowledge, the cognitive side of knowledge management*, Emerald, London.
Johannessen, J-A. (2022d). *The philosophy of tacit knowledge*, Emerald, London.
Johannessen, J-A. (2023). *Feudal capitalism in the innovation economy*, Routledge, London.
Johannessen, J-A. (2023a). *De-Globalization in the innovation economy:* ***Technological innovations trigger conflict between China and the United States*** Routledge, London.
Johannessen, J-A. (2024). *Safe jobs in the innovation economy: Strategic competence creation*, Routledge, London.
Johannessen, J-A. (2025). *Uncertain jobs in the innovation economy: Strategic competence destruction*, Routledge, London.
Joseph, W. (2019). *Politics in China: An introduction*, Oxford University Press, Oxford.
Kahan, S. (2013). *Getting innovation right: How to turn ideas into outcomes*, Jossey Bass, London.
Kotkin, J. (2020). *The coming of neo-feudalism: A warning to the global middle class*, Encounter Books, New York.
Krugman, P.R. (1995). *Peddling prosperity: Economic sense and nonsense in the age of diminished expectations*, Norton, New York.
Kurzweil, R. (2005). *The singularity is near*, Penguin, London.
Kurzweil, R. (2008). *The age of spiritual machines: When computers exceed human intelligence*, Penguin, London.
Kurzweil, R. (2013). *How to create a mind: The secret of human thought revealed*, Penguin Books, New York.
Mahbubani, K. (2018). *Has the west lost*, Allen Lane, New York.
Mahbubani, K. (2020). *Has China won?* PublicAffairs, London
Mann, G. (2019). *In the long run we all are dead: Keynesianism*, Political Economy and Revolution, Verso, London.
Marx, K. (1976). *Capital, Volume I*, Penguin Classics, London.
Marx, K. (1978). *Capital, Volume II*, Penguin Classics, London.
Marx, K. (1991). *Capital, Volume III*, Penguin Classics, London.
Mattingly, D.G. (2020). *The art of political control in China*, Cambridge University Press, Cambridge.
Milanovic, B. (2016). *Global inequality*, The Belknap Press, New York.
Oppenheimer, A. (2019). *The robots are coming: The future of jobs in the age of automation*, Vintage, London.
Picket, K. (2017). Foreword, in Brown, R. (ed.), *The inequality crises*, Policy Press, London, pp. vii–viii.
Piketty, T. (2014). *Capital in the twenty-first century*, The Belknap Press of Harvard University Press, Boston.
Piketty, T. (2016). *Chronicles: On our troubled times*, Viking, London.
Raworth, K. (2018). *Doughnut economics*, Random House, New York.

Reeves, R.V. (2018). *Dream Hoarders: How the American upper middle class is leaving everyone else in the dust. Why that is a problem and what to do about it*, Brooking Institution Press, Washington, D.C.
Rogers, E.M. (2003). *Diffusion of innovations*, Free Press, New York.
Schumpeter, J. (1951). *Theory of economic development*, Harvard University Press, Boston.
Schumpeter, J. (1954). *History of economic analysis*, Oxford University Press, Oxford.
Schumpeter, J. (1989). *Business cycles*, Porcupine Press, New York.
Schumpeter, J. (2002). Seventh chapter of the theory of economic development, translated by U. Backhaus, *Industry and Innovation*, 9: 93–145.
Schumpeter, J. (2006). *Theorie der Wirtschaftlichen Entwicklung*, Dunker & Humblot, Berlin.
Schumpeter, J. (2010). *Capitalism, socialism and democracy*, Routledge, London.
Schwab, K. (2016). *The fourth industrial revolution*, World Economic Forum, Geneva.
Schwab, K. (2018). *Shaping the fourth industrial revolution*, World Economic Forum, Geneva.
Skarzynski, P. & Gibson, R. (2008). *Innovation to the core*, Harvard Business School Press, Boston.
Standing, G. (2014). *The precariat: The new dangerous class*, Bloomsbury Academic, New York.
Standing, G. (2014a). *A precariat charter*, Bloomsbury, London.
Standing, G. (2016). *The corruption of capitalism: Why rentiers thrive and work does not pay*, Biteback Publishing, London.
Standing, G. (2019). *Plunder of the commons, a manifesto for sharing public wealth*, Pelican, New York.
Stilwell; F. (2019). *The political economy of inequality*, Polity, London.
Susskind, D. (2020). *A world without work: Technology, automation and how we should respond*, Allen Lane, London.
Susskind, J. (2018). *Living together in a world transformed by technology*, Oxford University Press, Oxford.
Thucydides. (2009). *The Peloponnesian war*, Oxford University Press, Oxford.
Tucker, R.B. (2002). *Driving growth through innovation: How leading firms are transforming their futures*, Berrett-Koehler Publisher, San Francisco.
Unterberg, B. (2013). *Crowdstorm: The future of ideas, innovation, and problem solving is collaboration*, John Willey & Sons, London.
Von Hippel, E. (2005). *Democratizing innovation*, MIT Press, Cambridge.
Zaltman, G., Duncan, R., & Holbeck, J. (1973). *Innovations and organizations*, Wiley, New York.

3 Artificial intelligence

Core idea in this chapter

The most significant consequence of robot shock is that most people will become serfs of their new feudal overlords, while continuing to believe that they have complete freedom.

Key points in this chapter

- In the innovation economy, power will accrue to those with access to large quantities of data.
- An evolutionary algorithm can be creative and rules-based at the same time.
- Intelligent robots apply an algorithmic rationality that is more effective than humans' bounded rationality.
- The operation of binary codes in an integrated network will generate a complexity that will give rise to new emergent systems, which are capable of developing creativity.
- The ability to understand and apply binary codes in practical operations will be crucial for success in the labour markets of the future.
- If everything is about being rational, then social systems will become irrational in their consequences.
- We act in light of the information in our possession. At the same time, knowledge transforms the knower. The practical effect of this is that the algorithms that we are exposed to will control our thinking and behaviour.
- Contextual competence and systemic insight will be important attributes for success in the innovation economy.
- Businesses that are unable to generate innovations, or to succeed in adapting to innovations generated by others, will become uncompetitive.

Introduction

In this chapter, we will examine the following question: What is the potential impact of artificial intelligence on future employment trends?

Artificial intelligence is an example of a general-purpose technology (GPT).[1] Other examples include steam power, electricity and information technology. Due

DOI: 10.4324/9781003567325-3

to their generic nature, GPTs have very wide ranges of application (Frey, 2019: 305). Historically, GPTs have tended to be adopted initially in a small number of specific areas before spreading throughout the economy. As a result, a GPT may have different applications, and also different consequences. These consequences are difficult to identify at the time the particular technology is adopted. There is one concept, however, that is often applied when a technology spreads throughout a social system. This concept is creative destruction (Aghion et al., 2021). The political economist Joseph Schumpeter was the first to apply this concept in an economic context. In various works, he demonstrated how creative destruction can influence the development of social systems.[2]

Algorithms are particularly important for the further development of artificial intelligence. We have also seen that the technology being developed is so multifaceted and of such small dimensions that it can be used in many industries. For instance, nanotechnology is one such technology that will have many areas of application in various industries. This development will lead to an increasing complexity at various system levels. It is also probable that future AI and nanotechnology applications will affect the development of our health services. We have already seen this technology transform banking and finance. In these industries, algorithms and nanotechnology solutions have had an impact on the buying and selling of shares and securities, as well as loan disbursements.

The synthesis of the developments we have already seen, and which we can imagine developing further, is that a completely new society is evolving. This new society, which we can already see the outlines of, will not have a linear development. In other words, something qualitatively new is now evolving. It is a society where human intelligence will be challenged by a completely new AI technology.

The field of artificial intelligence research can be traced as far back as to the 1950s. It is only today, approximately 70 years later, that this technology is challenging human intelligence. We stress the fact that we are talking about 'human intelligence' here and not 'human consciousness' (Harari, 2017, 2018). We have previously discussed whether artificial intelligence may be able to develop some form of 'consciousness' in the future (Johannessen, 2022c). We have also previously written about the ethical aspects of the use of artificial intelligence (Johannessen, 2021a). In addition, we have also described and analysed whether it is possible to use artificial intelligence creatively in order to generate innovations (Johannessen, 2022a). Moreover, we have discussed various aspects of artificial intelligence in relation to automation in several places; for example, in Chapter 2 of this book and elsewhere in other books (Johannessen, 2020a, 2020b). Consequently, in this chapter, we will only focus on how the use of artificial intelligence can affect the workplace of the future.

In his book, *The Sciences of the Artificial* (2019), the Nobel Prize winner in economic sciences, Herbert Simon, was perhaps the first person to examine artificial intelligence methodically.[3]

The future technological revolution that will result from the further development of artificial intelligence will affect institutions, power relations and workplaces.

Moreover, it will probably create tensions and conflicts in the global economy (Johannessen, 2021, 2022).

There is a basic premise behind the development of artificial intelligence in countries and regions, such as China, the US, Japan and Europe, which is that AI technology does not have emotional intelligence, nor is it able to express independently, personal opinions. This premise is straightforward to understand. The background for this assertion is even more straightforward. In the agricultural society, it was the ownership of land that determined power relations. In the industrial society, it was the ownership of capital and capital goods that determined power relations. In the innovation economy, knowledge, new technology and data will be central factors determining power relations. The development of artificial intelligence is controlled and driven forward by access to Big Data and learning algorithms, amongst other things. In this context, it is the large technology companies who are first in line when control of Big Data is to be determined. National governments also invest in Big Data. For instance, the Chinese Communist Party harvests and utilizes Big Data for various purposes. In other words, access to, and applications of, Big Data, is becoming a central factor for determining power relations in the innovation economy, something which Harari (2018), amongst others, has pointed out.

If we do not understand this basic premise, we will have great difficulty in understanding why tensions and conflicts develop at different levels in social systems. This applies to everything from the actions of the large technology companies (Harari, 2018) to the growing tensions between China and the US (Johannessen, 2021, 2022). Without understanding this basic premise, we will also have difficulty understanding how and why societies are under pressure due to the increasing emergence of feudal structures (Johannessen, 2023; Kotkin, 2020).

In many countries around the world today, we can see feudal structures emerging or what we may term feudal capitalism. This is evident in the growing trend to vote for 'strong men', that is, leaders who emphasize 'my country first', defending traditional family values and the glorification of so-called national 'histories'. The new applications of artificial intelligence is strengthening this development, because intelligent robots can carry out productive processes without decisions being questioned.

Artificial intelligence, gene-editing, Big Data, block chain technology and other technologies that drive the innovation economy forward have one thing in common: All these technologies are driven forward by the access to data. Whether this data is on a biological level or on a behavioural level plays less of a role. Access to data gives power over social systems. When this access to data is connected to social structures that are becoming more and more 'feudal', then liberal democracy is threatened. People in the West may think they choose freely and that they are able to make independent decisions. However, recent applications of 'Data Analytics[4] have been utilized to influence public opinion in several ways. With this new scenario, involving the increased application of AI and access to data, it might be said that we are moving towards 'A Brave New World'.[5]

Artificial intelligence has come a long way since the time when Herbert Simon first researched and wrote about this technology in the late 1960s (Simon, 2019). At the time, questions were asked about how 'artificial intelligence' should be categorized and whether it belonged within the realm of science or not. Today, artificial intelligence has reached a level of development where it is becoming clearer that the technology is to a great extent driven by access to data and where this access to data is also related to power structures. It is not difficult to understand that the people who owned land in the agricultural society also had power. It is also not difficult to understand that the capitalist in the industrial society who owned capital goods, such as production machines, had more power than the people operating the machines. However, it is perhaps more difficult to understand that the people in today's information society that have access to larger amounts of data are the ones with the 'power'. It is precisely this lack of understanding that makes it possible for some people to usurp power, creating a new scenario that is beginning to increasingly resemble the feudal society, where the lords had all the power. This lack of understanding of the relationship between access to data and power can result in feudal capitalism emerging and existing for as long as the feudal society did; in other words, a new thousand-year era of darkness. Our aim in writing this book, and which we have also written about in other books, is to create a greater awareness regarding the above phenomenon.[6] Of course, we are aware of the fact that our contribution is no more than a grain of sand in one of the bricks that is being laid to build the new 'cathedral'. Although we are just small pieces in a much larger world, we nevertheless have a responsibility to the future. Therefore, this book, and the other books we have referred to in the footnote, may be viewed as being a 'wake-up call'. If we do not 'wake up' and understand the importance of access to data for how society will be developed, then we are handing our future over to the new elite, the feudal lords, who will shape the future according to their own self-interest; that is, creating a future that serves them and their children best. The most significant consequence of 'robot shock' is that most people will become serfs of their new feudal overlords, while continuing to believe that they have complete freedom.

Media, Big Data, artificial intelligence, gene-editing, learning and evolutionary algorithms applied in various forms of intelligent robots, all contribute to people believing that they have greater freedom. In this context, it may be appropriate to quote Nietzsche (1968: 35), who writes that 'freedom is a concept utilized by the upper classes in order to retain power'. It may be said that this ability to control people through access to data is what is truly frightening about the new technology. If you can control people's behaviour, by making them believe that it is their own free choices which are behind their decisions, then the new technology is like a dream come true for those who aim to manipulate, control and rule people. This type of power, where the individual does not believe he/she is being controlled, is truly a quantum leap in the application of power and technology. In this context, the ability of authoritarian and autocratic regimes of the present and past to control people's behaviour pales in comparison of what will be possible in a neo-feudalistic society in the near future (Johannessen, 2023). We have already witnessed elements of this future controlling society in the various manipulations of American elections, when data analytic companies harvested data illegally, and used the data to provide analytical assistance to presidential candidates (Sønderholm et al., 2021: 119–139). Therefore, the use of Big Data and Data Analytics to influence people's opinions and change behaviour is something that is already occurring and is likely to be further developed in the future. In this way, applications of the new technology will not only impact the workplaces of the future, but they will also influence how we think and react to events and actions. Access to data may seem abstract to the man in the street, but it will have critical tangible consequences for the individual, organizations and society. We are facing a time where 'robot shock' will lay the foundation of a new thousand-year era of darkness, where feudal structures will organize our societies (Kotkin, 2020). The paradox is that we ourselves welcome this development, because we believe we are 'free' and can make 'free' choices, while in reality we are governed through complex technological influencing mechanisms. Having access to data creates the paradox that we become our own worst enemies. We are manipulated through various forms of information into actually believing that being ruled by a 'feudal' elite is in our own best interests. It is against this background that the new technology that is emerging is qualitatively different from previous technologies.

The new technology, such as artificial intelligence, Big Data and robotics, will be aligned with our way of thinking and acting. In other words, it is what we think and how we think and act based on our mindsets that the new technology will target. The technology is, of course, applied in other ways, such as to improve methods of production, distribution and consumption. The fact that the technology can be applied to increase production is essentially positive. On the other hand, the fact that the technology can also be used to influence our thinking and actions poses a danger. It may sound unimaginable and completely bizarre that a technology can influence how we think, communicate and act. However, we have already seen the beginnings of this development (Sønderholm et al., 2021: 119–139).

The Fourth Industrial Revolution will be characterized by quantum leaps in productivity growth due to the further application of new technologies. However, there are other aspects of the new technologies apart from productivity growth. The new technologies, such as artificial intelligence, algorithms, robotics and so on, are cognitive technologies, or 'thinking' technologies. These cognitive technologies will effectively influence our thoughts and actions. In other words, the application of these technologies, in combination with the access to data, can influence our behaviour and actions in many ways, from our shopping to the way we vote in democratic elections. Thus, the unregulated use of these cognitive technologies can pose a danger to our liberal democratic societies. This fear has been expressed by key figures such as Stephen Hawking, and many experts, such as Elon Musk, Peter Norvik and Stuart Russel, who signed an open letter describing the possible dangers associated with the unfettered future development of AI technology.[7] It is perhaps even more frightening that these cognitive technologies can take over our minds, take control of our thoughts, how we think, what we think and not least how we act. In other words, it may be possible that the future development of cognitive technologies will target the psychology of human thinking, gaining knowledge about how humans think (Minda, 2020). It is when the technology affects our way of thinking that we will have lost control over the new technology. If this introduction has one message, it is that artificial intelligence, and the associated cognitive technologies, will become embedded in our thinking, our values, our norms and our actions. This is the real consequence of 'robot shock'. This consequence goes far beyond the productive power of the new technologies and the dangers Hawking and others have pointed out. If we are correct in this assumption, then the feudal structures will indeed develop into feudal capitalism. Unfortunately, this feudal capitalism may be something we will have to live with for the next thousand years.

We are focusing here on artificial intelligence as a social phenomenon, not as a technological phenomenon. We are concerned with the social consequences of using artificial intelligence. This is also the focus of the open letter signed by Hawking and the other signatories. The letter points out that society can reap great potential benefits from artificial intelligence (AI), such as the potential of AI to eradicate disease and poverty.[8] However, the letter also called for research regarding the pitfalls of AI, that is, that the technology could become unsafe and uncontrollable.

Narratives related to artificial intelligence

Case letter 1: Artificial intelligence and hierarchical organizations

Our organizations are hierarchical and often have management systems that attempt to control the production process, regardless of the products or services that are delivered. These hierarchical organizations have a major drawback, and that is their inability to adapt quickly to changing circumstances. One reason for this is that the many people whose task is to develop strategies and future scenarios only include their own functions in these scenarios to a small extent. What would happen if a new technology could take over most of the functions of staffs and departments that are used to develop future solutions? This would be a cognitive technology that could replace many, if not all, of the people who have such functions in organizations. Does it seem reasonable that those people in organizations who are concerned with planning the organization's future will develop and implement a technology that results in them becoming redundant? It is unlikely that this will happen. Therefore, hierarchical organizations are not only slow in adapting to changed circumstances, but they do not change satisfactorily either. The rationale is that those in decision-making positions are not interested in putting themselves and their close network contacts at risk. In this way, over time, tensions can develop between decision-makers and the people who are eliminated by the automation processes that result from the application of new technology. It is easier to automate away employees that have routine tasks and who are not closely connected to the decision-making processes than it is to automate away people who are concerned with decision-making. This assumption seems to be correct, and there are many indications that this is also the case regarding the development of artificial intelligence in relation to hierarchical organizations.[9] To put it a bit more bluntly, we can say that organizations have developed a 'fat layer' in the form of administrative positions, so they resemble more a star-shaped figure than a hierarchy. The star has emerged because the middle of the hierarchy has grown out to the sides so that the hierarchy has been padded with an administrative layer of fat. This fat layer prevents signals from the front line from quickly reaching the decision-making strata at the top of the organization. Once the decisions have been made, they are dampened on the way down by the administrative fat layer, so that they do not damage the fat layer consisting of management and administration. In this process, what the organization is designed to do becomes less important. The management and administration find new purposes for the organization, so that tensions are created between what the organization is designed to do and what it actually does.

When cognitive technology exists, but it is not put to use, then it is only a matter of time before completely new organizations emerge, which are not burdened with the historical anchors like the old organizations are. We have seen countless examples of such changes throughout history. For instance, if we consider the transition that occurred in the photography industry, with the demise of Kodak and the meteoric rise of Instagram. Instagram, using a digital business model, with only 13 employees, managed to outperform Kodak with tens of thousands of employees.

The question that can be asked is why an organization as big as Kodak didn't invent 'Instagram' themselves? The technology was available, and it was quite obvious that the decision to use this new technology would increase productivity, reduce costs and increase income for Kodak. Kodak did not make this decision. It was left up to some entrepreneurs from outside Kodak to introduce the new system, from which the customers greatly benefitted. The point in this context is not to describe the Kodak–Instagram case, but to suggest that Kodak had a decision-making structure, which failed to implement a technology that would result in many employees in the administrative layer losing their jobs. This is of course not difficult to imagine, and the person who has perhaps researched and written most about this phenomenon is the Nobel Prize winner in economics, Daniel Kahneman.[10] Kahneman and Tversky developed the Prospect Theory about why people resist change and what can be done to counteract this phenomenon (Adriaenssen & Johannessen, 2016a).

This small case letter says something about why it is difficult to implement effective AI technology in hierarchical organizations.

Case letter 2: Intelligent robots taking over jobs

Intelligent robots have already begun to automate work processes. This often occurs using computers that are not located in the same location as the actual work processes. This is termed Robot Process Automation (RPA),[11] which involves business process automation based on robots. In other words, it is a way of automating parts of the paperwork and office work.[12] These intelligent automation processes often merge with what can be called cognitive automation. Combining these two types of intelligent automation processes, RPA and cognitive automation, the new technology has started to automate the decision-making processes and the administrative functions of organizations in order to improve organizational outcomes. In other words, this development may be said to represent a watershed or turning point in the development of automation processes. If the operations of a business or public organization are divided into two parts, the part that produces goods and services, and the part that focuses on administrative processes, then new automation processes will not only automate production but also the administrative work processes.

RPA technology allows you to automate repetitive administrative work. This concerns the traditional division of paperwork into smaller parts so that you can separate a larger work task into smaller parts that are connected to each other. If-then procedures are then developed. The software is then able to deal with the automation of these operations. In the initial phase of such automation procedures, it is rare that the human workforce is affected in any other way than that they have to participate in turning routine tasks into if-then procedures. A little further on in this automation process, more people become redundant because at this point the software can carry out what the humans carried out previously. Viewed against this background, RPA is a process that first makes use of the existing human workforce, but then, at a later stage, more and more of those who previously operated these processes become redundant.

There are many examples of the use of RPA. For instance, RPA technology is used in travel reimbursements and for the claims processing and invoicing of insurance companies. There are also variants of the RPA system that are used in many 'chat' systems, where you can ask questions online, and then get answers from software programs regarding simple requests, such as closing times, opening times, the returning of goods and so on. These systems are also used to automate tax and financial operations.

RPA programs result in cost savings and faster processing so that customers receive more effective feedback. Regarding the use of RPA, a productivity increase of up to 70 per cent has been reported by KPMG.[13]

Cognitive automation is a type of RPA system that uses artificial intelligence. The largest difference to regular RPA programs is that cognitive automation uses algorithms so it is able to 'learn'. It is the use of learning algorithms that will very probably result in much progress being made in the automation of information, communication and knowledge processes of businesses and organizations. This will also concern the use of programs that are able to produce decisions far more complex than the automation of simple routine procedures. It is in relation to the application of these cognitive automation programs that human labour will be affected. In other words, it can be said that it is the cognitive automation programs that will very likely influence the development of the workplace of the future.

Description related to artificial intelligence

The Fourth Industrial Revolution[14] will be driven forward by technological innovations, specifically related to the core technology, artificial intelligence (AI). Other technologies associated with AI, such as Big Data, gene-editing, block chain technology and so on, will also drive forward the Fourth Industrial Revolution.

The new technologies that will be further developed during the Fourth Industrial Revolution will have the potential to eradicate poverty, hunger, and diseases and to improve people's well-being. On the other hand, how the new technology is applied will greatly depend on decisions made within political systems. In other words, whether people's well-being will be improved or not, based on the potential of technological developments, will depend on the political will of governments and authorities to achieve such an aim.

The new technology will have the potential to promote individual freedom. The question that arises, however, is how the individual will cope with this increased freedom and how the political system handles this opportunity for increased freedom.

Against this background, the Fourth Industrial Revolution will be characterized by an interaction between the technological system and the political system. Artificial intelligence will constitute the core technology of the technological system. The political system may be related to a system that enables greater freedom for more and more people. However, 'freedom' in this context is not necessarily about promoting the individual's needs and wishes in isolation. On the contrary, 'freedom' is not about the single individual but about living together as a strong community. An appropriate metaphor here is a bundle of sticks. Sticks in a bundle (community) can't be broken, but a single stick (the individual) can easily be broken. The same applies to people. There is strength in union. Thus, it is in union and community where the individual can find strength and 'freedom'. Artificial intelligence can also be understood in relation to this metaphor, as the technology that binds together the many other technological elements and the political system. However, how this social system will take shape will be determined by the political system.

Artificial intelligence can be related to many other fields and areas of expertise. In the following, we will focus on five elements: Algorithms, small-scaled robotics and nanotechnology, complexity, creative destruction and the industries of the future.

Algorithms

Algorithms can be 'dumb', they can be 'learning' algorithms, they can be 'evolutionary' or constructed in other ways. Robots with fixed and static algorithms will not be able to learn from situations. However, the new algorithms (evolutionary and genetic algorithms) will most probably be designed so that they can modify themselves after feedback from a situation, i.e., learning and then modification. This is not dissimilar to how a human being learns and modifies his/her behaviour. However, there is much to suggest that robots will be able to modify their behaviour much quicker than humans, because they will not be hindered by the many psychological mechanisms that can prevent humans from changing behaviour.[15] However, we do not find these processes that inhibit change in an intelligent robot.

In an intelligent robot, genetic algorithms[16] can be modified when they receive feedback from a situation. The human gene pool, on the other hand, only changes after generations.[17] Intelligent robots with integrated genetic algorithms will be able to change their behaviour in a few nano-seconds. In this way, one can assume that the intelligent robot will be able to utilize self-reflection and modify its behaviour very much quicker than any human (Mishra et al., 2019).

Genetic algorithms attempt to mimic natural selection and the evolutionary process and belong to a group of algorithms called evolutionary algorithms (Zhou et al., 2019).

If we think of algorithms as a type of hierarchical decision tree, with its if-then procedures, then we have not understood how the many different types of algorithms are changing work processes that are cognitive and not purely routine tasks. It is the new algorithms, the learning and evolutionary ones, that will lead to the workplace of the future being qualitatively different from today's workplace.

Using evolutionary algorithms, intelligent robots will be able to evolve in analogy with the learning process undergone by humans, only at a rate that is much more rapid than in humans. Some robots learn and adapt to the world faster than other robots, because their basic designs in relation to the algorithms are different, not unlike human genes. We will be able to further develop the robots that learn and adapt quickly, which will lead to the creation of a new type of robot, the evolutionary robot. Once these evolutionary robots have been developed, singularity will have become a fact; this is expected to occur around 2035–2040 (Kurzweil, 2005, 2008, 2013). Singularity[18] in this context means the point at which machine intelligence surpasses human intelligence. We are not there yet, and it may take a long time before we get there. The point, however, is that in today's situation, where everyone is struggling with both new technology and understanding how the new technology will develop in the future, most people use some form of algorithms and intelligent robots in their workplaces. With such an understanding, it is only a matter of time before the learning algorithms take over more and more of our work tasks. This is not to say that we are approaching mass unemployment. It is more the case that new tasks and new competences will be in demand in the future. An important point we discuss and analyse in Johannessen (2022d, 2024, 2025) is that the effects of automation resulting from the deployment of artificial intelligence and intelligent robots are not visible in the unemployment statistics but in the wage statistics.

We are used to thinking that machines cannot be creative. It is, therefore, important to be aware of the fact that there is a separate process in an intelligent robot that stores and retrieves information. This memory unit operates not only on a logical-rational level but also on a creative level. The creative level can be imagined in the following way. The information that is extracted from the memory unit is done by algorithms that have been developed to be innovative and creative and not only logical and rational (Johannessen, 2022a, 2022c). This can work in the following way. On the one hand, the memory unit is based on algorithms that instruct the program to obey some logical rules. On the other hand, the instructions in the algorithm are also 'illogical', so that the creative and new, through various combinations, can emerge from the memory unit of the intelligent robot. Memory understood in this way is two-sided, logical and illogical at the same time, but of course following different time sequences. We can only imagine what such a development will mean for the workplaces of the future.

Small-scaled robotics and nanotechnology

Links between small-scale robotics and artificial intelligence are well illustrated by the following example of technological innovation, described in a newspaper article about the work of Chinese researchers published on 26 May 2022.[19]

The article describes how Chinese researchers have developed a miniature robot with 'muscles' and 'feet'. This tiny robot can wriggle in earthworm-like fashion through pipes of less than a centimetre in diameter. The robot, which weighs 2.2 grams and is 47 millimetres in length, is electrically powered by means of a tether. Potential areas of application for this tiny robot are endless. For example, it can be used to inspect complex, narrow pipelines inside engines. Much smaller robots, known as 'nanobots', can be used to clear clogged arteries.

A general search of the journal 'Science Robotics' for 'nanorobots' (6 October 2022) generated 10,958 hits. This number indicates that the development of nanotechnology is well underway. Accordingly, it is only a question of time until this technology is applied in a multitude of highly complex operations. In particular, as mentioned above, in the future, it may be possible to use nanotechnology to treat artery disease.

As also mentioned above, the researchers from Tsinghua University in China used a tiny robot with 'muscles' and 'feet' that can propel its way through pipes for inspection purposes. The robot can propel itself both horizontally and vertically at a rate exceeding its body length per second.[20] The researchers found that the robot can also propel itself through pipes of varying geometries, such as L-shaped and spiral pipes with variable diameters, as well as through pipes filled with either air or oil. The researchers used magnets to assemble the robot's modular components so that it can pass with ease through pipelines of varying shapes.

In addition, the robot has a soft body that allows it to adapt to different pipework conditions. The point of including this technical description is to show how in the future, micro- and nanorobots will most likely become vital components in many different kinds of instruments, with medical instruments being just one example. This becomes particularly apparent when we read that the researchers attached a small endoscopic camera to the tiny robot, allowing pipework inspections to proceed at different speeds.

When Alec Ross (2016) wrote his book, *The Industries of the Future*, the field of small-scale robotics was in its infancy. In 2022, this technology has progressed so far that researchers are developing micro- and nanorobots for use in inspecting the interiors of complex pipework systems. Ross described how silicon had replaced aluminium in the production of miniature robots. Today, we see the rapid development of completely novel materials for use in micro- and nanorobotics. These materials can be used to make flexible components that can function as artificial muscles to propel the robot on its inspection missions. Once we are aware of these developments, the question is: In the future, what parts of the labour market will not be affected by micro- and nanotechnology? Most jobs will be affected in one way or another by this technology.

There is much to suggest that there are three technologies that will have a major impact on jobs in the future: biotech, information technology and nanotechnology (Kumar Singla et al., 2021: 1). In particular, we can expect these innovative technologies to have significant practical consequences in the areas where artificial intelligence and nanotechnology converge. This is the area that will have an impact on many highly complex sectors, including healthcare, agriculture, engineering science and geology.

As far as healthcare at an individual level is concerned, the potential implications of nanotechnology combined with fields such as artificial intelligence, information science and cognitive science are scarcely imaginable. Previously, our bodies have largely been viewed as a 'given', but now we can start to imagine how problems associated with ageing could be ameliorated or eliminated by the application of new technology. If such a future materializes, and many people are claiming that it will,[21] then our workplaces will be transformed in ways that are more far-reaching than changes to work processes. We can imagine how the skills that an older person has built up over the course of a long career could be used for longer than is the case today.

Complexity

There appears to be a particular phenomenon that emerges in relation to integrated computers and to the development and application of artificial intelligence. This phenomenon is complexity. The complexity is found in the computers and in intelligent robots that use learning algorithms, as well as in the environment in which these intelligent robots operate. This complexity is then used in many cases to further develop artificial intelligence. We see this quite clearly in the application of Big Data and facial recognition. In both of these two cases, technology is used to better understand human behaviour, and in some cases, this understanding of human behaviour is also used to control behaviour. If we use Big Data to understand human behaviour, then those who are interested in making use of this knowledge can, in many cases, know what the individual will do in various situations and contexts long before the individual himself knows it. This can apply to purchasing behaviour, voting behaviour and other situations where there is a large degree of complexity in the choices of action.

Behavioural psychology and neuro-psychology will greatly benefit from the insights that can be gained from the use of Big Data, facial recognition and other data collection methods in relation to understanding human behaviour. In this way, artificial intelligence can be used when the complexity is great in order to develop professional knowledge in fields such as psychology, sociology and possibly also in international relations.

Another field that will be able to benefit from the development of artificial intelligence is economics. Decision-making in economic systems often has to deal with complex information situations. The global economy is becoming increasingly complex and characterized by multifaceted information processes and many economic actors. This connection between the many global actors and the great complexity of information processes will necessarily lead to problems for human decision-making. Using artificial intelligence, this complexity can be reduced to effortless decision-making, because the smart robots can make rational and efficient decisions using complex datasets. The smart robots are able to do this, because they have knowledge of how groups of economic actors and individuals will react to various changes in economic variables. These machines know this because they have simulated various events millions of times in the same way that chess computers have played against themselves millions of times in order to find the best moves. In other words, a high degree of complexity may pose a disadvantage for the human decision-maker, but this is not the case for smart robots. If we assume that increasing complexity is a growing phenomenon in the global economy and global relations, then this will very probably promote the development and application of artificial intelligence. The rationale is that this technology will reduce complexity and make decision-making processes more unproblematic for decision-makers. This development suggests that a new form of rationality will emerge. Smart robots with AI algorithms integrated into them will be able to create a new kind of rationality, 'algorithmic rationality', which does not have the same limitations as the 'bounded rationality'[22] of humans, which often results in decisions not being optimal. In this way, it can be said that the new technology will replace the merely adequate decisions made by humans, with the optimal decisions made by smart robots. Whether this will benefit social systems is another question, which we will further elaborate on below.

The proposition that can be derived from what we have described so far about complexity is the following: Limited human rationality, when dealing with a large degree of complexity, can be offset by using smart robots.

Creative destruction

Increasing complexity promotes the development of creative destruction. In other words, this is when the destruction of 'the old way' of doing things is replaced by 'the new way' of doing things. This can also be understood as one crisis following another. This is related to the fact that when the old is destroyed, many people and businesses will no longer have the necessary skills to compete in the new market. On the other hand, these crises are only a temporary phenomenon. It is the time it takes for the social system to adapt and develop new skills which results in crises. When the competence that is compatible with the new technology is developed, the crises will be defused. At this point in time, the social system will have entered a creative phase, where the new skills are used to develop new jobs and new businesses.

The above description of creative destruction is linear. However, in industrial revolutions, such as the imminent Fourth Industrial Revolution, these destructive tendencies come in cascades. In other words, it is not the case that one crisis will appear and then move into a creative phase. The various technologies will impact the market like a big wave crashing on the beach. During an industrial revolution there will be thousands of such big waves crashing on the beach one after the other. Before one wave has completely subsided, a new one will follow in its path.

The core technology of the Fourth Industrial Revolution is artificial intelligence. Related to this core technology are other technologies, such as Big Data, gene-editing, data storage, cognitive technology, blockchain technology and so on. When the various technologies interact, businesses will experience increasing complexity and continuous changes. These continuous changes create a type of erratic behaviour in businesses, that is, before a business has had time to adapt to one change, another change appears. Such erratic behaviour reduces the efficiency and productivity of businesses. This is why it is crucial for businesses that are impacted by cascades of changes that they learn how to manage complexity. We will elaborate on this further below, in the 'analysis', 'theoretical reflections', 'review of the practical utility' and the 'systemic aspects'.

Creative destruction that occurs in an industrial revolution will very likely lead to economic shocks entering social systems at all levels. These are innovations that continuously enter the market. These innovations lead to the development of large differences in productivity between businesses, regions and nations. It is those businesses that first implement the new innovations that will be able to increase their productivity. When these businesses increase their productivity, they will be able to attract investors and people with the best skills, because revenues increase, while costs are reduced. Those businesses that do not implement the new technology will end up in an economic backwater. Therefore, innovations will lead to some social systems benefitting in relation to others. This increased productivity and profit will then be reinforced by the fact that the business in question can purchase additional new technology and expertise to further increase its productivity. In this way, 'first-rate and second-rate players and teams' will emerge in social systems. This will become evident in the contrast between urban and rural areas, between the highly skilled and those with few skills, between the elderly and the young and between the various regions and nations. Against this background, the following propositions have been developed:

Proposition: Minor innovations lead to minor economic crises.
Proposition: Major innovations lead to major economic crises.

These propositions are another way of describing Schumpeter's (1954, 1989) concept of creative destruction.

Industries of the future

The Fourth Industrial Revolution will be an arena for dealing with an increasing amount of data, information, communication and expertise. Information and communication technology is under continual development, and new forms of technology are being used in our workplaces. This new technology is mainly based on artificial intelligence, and machine learning has now reached a level where these machines are making decisions previously made by humans. For instance, artificial intelligence is now being used when people apply for bank loans. To an increasing extent, this is done online. For mortgage lending, algorithmic lending procedures determine if loans should be granted in relation to credit scoring models. The borrower rarely has the opportunity to come into contact with a human assessor (Zuboff, 2019). In this way, artificial intelligence has already become important in work functions in our workplaces (Kailia, 2016: 9; Jones, 2021).

Artificial intelligence and intelligent robots are the new 'employees' that human resource departments need to deal with. We have to learn to live with these intelligent robots in our workplaces, says Alec Ross (2016). There are already several types of intelligent robots that have taken over many work processes in many branches, such as in stock analysis, banking transactions, e-commerce, pharmacy, healthcare and so on. In many of these branches, work processes are characterized by human-machine interaction. However, in the future, it is very probable that in many cases the intelligent robots will be further developed and will be able to work independently and not controlled and operated by humans (McDonagh, 2021; Delfanti, 2021). The new technology that is being increasingly used in work operations is based on a coding system that uses the binary digits 0 and 1. Although a binary code is the starting point for the application of the new technology, the consequences of using such applications are not 'binary'. When these binary codes operate in an integrated network, they will generate a complexity that will give rise to new emergent systems.

Being able to understand and apply these binary codes in practical operations will be a necessary, if not sufficient, prerequisite for succeeding in the future labour market. However, this should not be understood as meaning that the industries of the future will consist of 'human robots' who are only concerned with codes and algorithms. It is not possible in the present day to develop algorithms for trust, friendship, network understanding, cooperation and all the factors that make us human. Being loyal, tenacious, inspiring confidence and having a good mood that is infectious will always be qualities that are sought after. This is why we also say that the technological competence in the industries of the future is only a necessary prerequisite. The sufficient prerequisite are those human qualities that make people into trustworthy and loyal employees. Having said this, skills in the development and application of coding in relation to work functions will secure the individual's future. This may sound like skills that only a few people will be able to acquire. However, even in those workplaces where there is a need for hands-on staff that can take care of customers and users, skills in coding will also be decisive for the quality of work performed. This is evident in several professions; for example, psychologists and doctors can use coding to transform written information about a patient's health condition, such as medical procedures, diagnosis and so on. This enables them to increase their productivity and efficiency.

The people who owned land in the agricultural society were the ones with wealth and power. The capitalists in the industrial society who owned capital goods, such

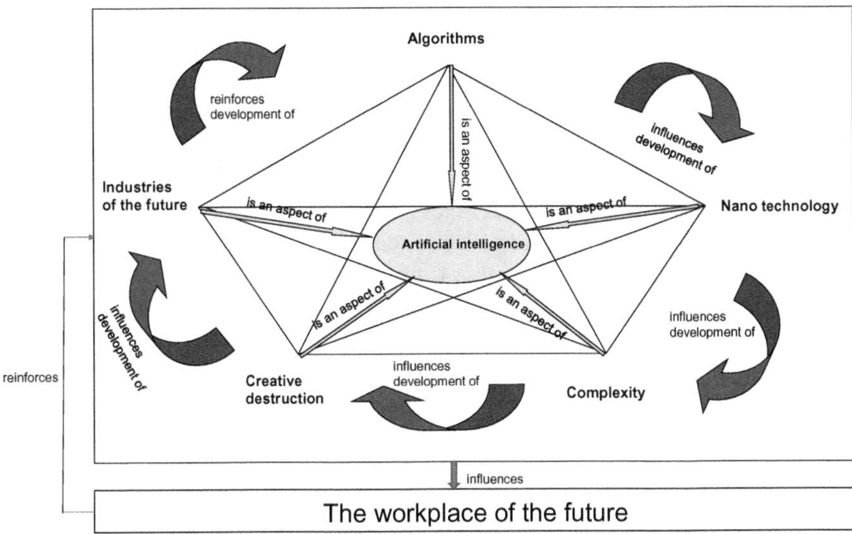

Figure 3.1 Aspects of artificial intelligence in the workplace of the future: A conceptual model.

as production machines, were the ones with wealth and power. In tomorrow's Fourth Industrial Revolution, those people who have access to large amounts of data and who are able to apply AI to process this data will also be the ones with wealth and power. In other words, skills and expertise in AI will become the most important competence in the innovation economy.

Based on the above description, we have developed the following proposition:

Proposition: Data, coding and artificial intelligence will change the workplace of the future, like the motions of tectonic plates can result in significant changes in the Earth's geography.

Based on the above description, we have developed a conceptual model, which combines the five elements that have been discussed (Figure 3.1).

Analysis related to artificial intelligence

Intelligent robots apply an algorithmic rationality that is more effective than humans' bounded rationality (Simon, 2019). The unlimited rationality of artificial intelligence and the limited rationality of humans is also the distinction that makes humans human. The emotional, social and relational elements show that where limited rationality prevails in decisions, there is also room for the aspects of what it means to be human. Being human shows itself in everything that is not fully rational. However, this is not synonymous with irrationality. The limited rationality is supplemented with the other elements mentioned above, which are not necessarily related to so-called logical rationality. In this way, we can say that the workplace of the future may not be as rational as it could have been. However, it will be more humane if it does not completely rely on algorithmic rationality.

It is precisely the complexity of the world around us that means that it will not always be the best solution to only use algorithmic rationality in decision-making processes. On the other hand, there is a current trend to view data as the new 'gold'. When, not if, data becomes as valuable and important as land was in the agricultural society and machines were in the industrial society, then the complexity of the world around us will increase. When this occurs, people and social systems will need to develop a form of requisite variety, so they can deal with the increased complexity. It is in such a situation that social systems must resort to further use of computing power to deal with the increased complexity. This computing power is related to the applications of algorithms and artificial intelligence, amongst other things. When this technology becomes an important feature of our workplaces, this will also result in increased complexity. It is in this context one can say that data will be the new 'gold' for businesses, stimulating production and fuelling innovation, as well as increasing complexity. This increase in complexity will be reduced by businesses and other social systems by developing solutions to deal with the increasing variation. The end result for many employees will be that they will experience feelings of powerlessness and alienation, because intelligent robots take over work processes and functions that were previously done by them.

The limited rationality of people has had 'positive' consequences in the past and present, because the 'human' aspect could be emphasized; however, algorithmic rationality is now beginning to increasingly dominate work processes. Data will be the most important raw material in the Fourth Industrial Revolution; coupled with the possibility of using this data in conjunction with new technology, such as artificial intelligence, this will probably result in the workplaces of the future moving more and more towards a situation of 'total rationality'. This may sound efficient and productive. However, an important aspect of this total rationality is the limitations for the people functioning in these social systems. Their emotional, social and relational qualities will of necessity be given less emphasis, when everything is about being 'rational'. In other words, if everything is about being rational, then social systems will become irrational in their consequences.

Our behaviour is largely adapted to the goals we set for ourselves and the social system we are part of. If our behaviour is dependent on the circumstances and developments in the world around us, our goals, and the goals of the social system we are part of, can come into conflict. In order to mitigate this conflict, we may need to adopt a technology that can facilitate greater requisite variety in relation to the outside world. Such a development will increase the pressure on our behaviour and possibly also our goals. To cope with this pressure, we will need to acquire the necessary technology to develop our requisite variety. This develops into a paradox, because we know that the complexity of the world around us makes new technology necessary. We also know that if we are to cope with this complexity, we must also gain access to this technology. When we gain access to this technology, we also know that it will lead to us letting go of the positive aspects of our limited rationality. In this way, we use algorithmic rationality as a possible solution. In its consequences, this leads to us using a technology that transforms us into instruments of algorithmic rationality. Thus, we become what we do not want to become.

The proposition that we can deduce from the above analysis is the following: Algorithmic rationality will change our workplaces so that limited rationality will be less in demand.

Theoretical reflections related to artificial intelligence

In this section we will propose the following proposition: When the degree of application of algorithms in work processes is great and the degree of complexity in the same work processes is great, then the probability is great that there will be an increased application of artificial intelligence in the work processes of the future.

An important challenge for the work processes of the future is that they will be increasingly dependent on data from the outside world in order to function effectively. This can be explained by the increasing interconnection at all levels of global data and information processes. This increasing interconnection leads to increasing complexity. Another aspect of this interconnection is that there will always be some areas that are interconnected while other areas are not. Metaphorically, this may be compared to a situation where all the people at a gathering talk to everyone else who is there, but no single person would have an overall understanding of what had been said by everybody. Similarly, on the global stage, all the social systems cannot simultaneously be in contact with each other. There are various separation mechanisms in operation regarding the interconnections in the global economy, resulting in a certain structuring.

When many people are connected to information that is available, this affects how people acquire information and how the information is used. Various platforms, such as Google, can be used to gain access to this information. The algorithms developed by Google enable people to gain access to some types and classes of information, while giving less access to other types and classes of information. In this way, the various platforms function as gatekeepers to various types of information. Regardless of how these platforms develop their algorithms, the algorithms are designed in such a way so as to provide access to certain types and classes of information, while giving less access to other types and classes. In other words, regarding access to the net, what we know is to a great extent governed by the various algorithms that are in operation. Moreover, we act in light of the information that is in our possession. At the same time, knowledge transforms the knower. The practical effect of this is that the algorithms that we are exposed to will control our thinking and behaviour. The greater the degree to which algorithms are used in work processes, the greater the probability that artificial intelligence will become part of these work processes, and consequently greatly influence our behaviour.

Our smartphones give us access to all the world's factual knowledge, both current and historical. With a few simple keystrokes, we can ask questions and receive answers. Translation software can translate written text or oral input into other languages and give written and oral responses. Algorithms are used in these processes in order to reduce complexity. In addition to searching for information, we can also communicate with other people, in every imaginable way, and at all times of the day. This way of communicating is viewed as increasing complexity by some, but it can also be perceived as a way of reducing complexity. For example, a salesperson representing a company in a country several thousand miles away may receive a question from a customer, about which the salesperson has no knowledge; the salesperson may be able to quickly answer the question after gaining access to others or sources that provide information about the question. This simplifies the process for both the salesperson, the customer and the business. On the other hand, there can also be many examples where such a connection can greatly increase complexity.

Digital technology has become possible through increased competence, the development of various forms of networks, as well as an agreement on some basic forms of standards. Without this agreement on standards, any interconnection would be almost impossible. For example, if each country has its own standard regarding railroad gauges, then it would become impractical to transport goods and people between the various countries by rail. Similarly, in the digital world, if no standards are agreed on regarding the transmission and storage of data and information, then the complexity would increase so much that it would become inefficient to use such digital systems. On the other hand, those countries that first adopt their own standards will be able to control the development of digital technology and the digital economy to a greater extent than the countries that have to adapt.

It is precisely this struggle for standards that can explain why China implemented the project, 'China Standards 2035'.[23] The preceding project 'Made in China 2025'[24] was a plan to transform China from being 'the world's factory', with the export of cheap goods for Western consumers, to becoming a leader in high-tech production. This project has been so successful that the US passed a law prohibiting the export of high-tech equipment to China. The American strategy has been to get other 'friendly' nations to join in this containment of Chinese high technology.[25] This was mainly technology related to the production of microchip technology. 'China Standards 2035' has had a slightly different aim than 'Made in China 2025'. The aim of 'China Standards 2035' is to develop future technology based on standards developed in China. These standards relate to the development of 5G and 6G networks, as well as the Internet of Things (IoT), artificial intelligence, quantum computing and biotechnology.

'China Standards 2035' focuses on standardization within 33 areas, related to the production of high-tech industrial goods, information and communication technology, the service industry, the carbon industry, the ecosystem industry and also within industrial development in rural areas.

The plan commits Chinese businesses to participate in the standardization. This is done through financial incentives if they follow the set standardization procedures. When enough businesses and entire industries have converted to the Chinese standards, they can then connect seamlessly, so that both efficiency and increased productivity can be achieved. Western businesses that do not use Chinese standards, and that want to cooperate with China, will face competitive challenges. One can only imagine the competitive conditions for German car companies that want to export their cars to China, and that do not adapt to Chinese standards. Against this background, it is important to be aware of the fact that in 2021, China became the largest market for German cars.

One can only imagine how the tensions between the US and China will increase within technological competition when we are aware of what is happening in relation to standardization. One of the consequences of the Chinese standardization process is that the technological innovations in China will be focused around these standards. This means that all new technological developments in China, e.g., in the field of artificial intelligence, will have the Chinese standards as their pivot point. If such a development becomes a reality, and much suggests this, then this will very likely increase the technological competition between the US and China. For the individual businesses in the West, e.g., the car industry in Germany, which has China as its largest market, the complexity will increase.

The standardization initiatives of 'China Standards 2035' will affect both businesses and nations. Businesses that have relations with China will increase their complexity in the production process. The US and its allies will find themselves in an increasing technology competition with China. This will very likely increase tensions between China and its allies on the one hand, and the US and its allies on the other. Given this development, the workplace of the future will become more complex. If China manages to dominate the development of standards for its high-tech production, as well as influence the development of standards in those businesses in the West that depend on exports to China, then China will manage to take control of high-tech processes in some Western countries, as well as all the businesses that wish to export to China.

Based on the above narratives, description and theoretical reflections, we have developed a typology for the application of artificial intelligence in the workplace of the future (Figure 3.2).

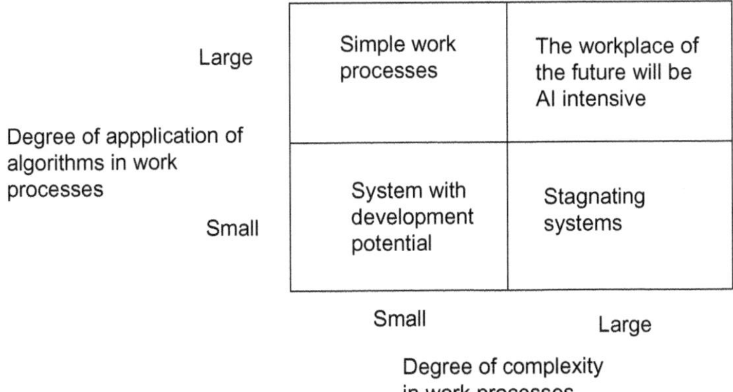

Figure 3.2 Algorithms and complexity related to AI in the workplace of the future: A typology.

Practical utility related to artificial intelligence

As the knowledge related to intelligent robots reaches a larger number of workplaces, this will increase not only the internal complexity of the workplaces but also the complexity between the individual businesses. The rationale is that more and more systems will come into contact with each other, and decisions will increasingly be made by intelligent robots. The knowledge developed through these interactions will be stored in various ways and be easily accessed by the intelligent robots. When these intelligent robots are connected in global networks, we term them 'intelligent informats'. We have done this to make a distinction between intelligent robots that function independently in workplaces and intelligent robots that are connected to other similar robots in the global space. Surgical robots are one such example of intelligent informats. These surgical robots are connected to other similar robots in the global space and can be constantly updated with new knowledge about the operations that the robots perform. This updating in real-time requires efficient and robust global networks, such as 5G networks and later 6G networks. 6G networks will be able to use higher frequencies than 5G networks and provide considerably higher capacity; in addition, 6G will be genuinely artificial intelligence-driven communication. In other words, the connections between intelligent robots, intelligent informats and artificial intelligence will create increasing complexity in work processes. However, those people working with these processes will not be aware of the increasing complexity, because all the decisions will be made by the various robots. On the other hand, one can conceive that when all or most decisions are made by intelligent robots and intelligent informats, then people in their workplaces will feel more powerless in relation to their own work and perhaps experience feelings of alienation and a growing sense of meaninglessness regarding their work.

When intelligent robots and intelligent informats become a normal component of most workplaces, it is very probable people will eventually adapt to the new working conditions. In other words, people's feelings of powerlessness and alienation regarding their work will probably be a transitional phenomenon. Workers in the past have also adapted to new working conditions, so this is nothing new. During the previous industrial revolutions, people had to adapt to new technology, such as the steam engine, electricity, automobiles, trains, computers and so on. In this context, the 'robot shock' will only constitute a transitory phenomenon. In other words, when people have adapted to the new situation where intelligent robots and artificial intelligence are integral components of working life, they will no longer feel frustrated by the fact that their workplaces have changed qualitatively. As we have pointed out many times before, time lag is the most important factor in this process. However, in this context, one might be tempted to quote J. M. Keynes when he ironically noted, 'In the long run we are all dead'; in other words, people are more concerned about their present life than the future.[26]

In other words, when the individual worker is exposed to these changes in the here and now, then he/she will have to deal with them. This is where one can truly say that 'robot shock' will constitute a long-term situation for the individual worker. One can, of course, argue that after a passage of time, when the new generation takes over the workplaces, equipped with the appropriate education, so they can adapt to the new technology, they will not experience feelings of powerlessness or alienation. This is where the relevance of time lag comes into play.

The importance of time lag is evident if we examine the first three industrial revolutions. It took a long time before the new technologies of the First Industrial Revolution (1750–1840) impacted all the areas of industry and the economy. Some critics suggest it took more than a generation. The 'new' generation were better equipped to adapt to the new reality in workplaces. However, it took a shorter time for the technologies of the second (science and mass production) and third (digital technology) industrial revolutions to impact all areas of society. In other words, with each new industrial revolution the time lag has been reduced from when new technology is introduced to the point when it fully impacts the economy. If this assumption is correct, the new technologies of the imminent Fourth Industrial Revolution will take less time to penetrate work processes; this will mean that workers will have less time to adapt to the new technologies.

The future development of new technology, especially artificial intelligence and robotics, will lead to increased complexity in working life. Moreover, some skills will become redundant, and some businesses will fail to adapt to the new technology. At the same time, as this destruction takes place, new areas of competence will be created which will result in the emergence of new businesses. It is this process that we term creative destruction. We have witnessed such processes throughout economic history, especially in relation to the various industrial revolutions. However, it is not only the industrial revolutions that promote creative destruction. One could argue that creative destruction happens all the time whenever new innovations enter the market. In this context, one can say that minor innovations lead to limited creative destruction, while major innovations lead to a large extent of creative destruction. Put another way, throughout economic history, new innovations have resulted in some established businesses not being able to adapt, resulting in falling profits or bankruptcies. A good example of this phenomenon is the Swedish company Facit. They had the highest productivity in the production of mechanical calculating machines but ignored the new electronic calculating machines which entered the market. The end result was that Facit went into liquidation even though they had had the market's highest productivity. This is a classic example of businesses that do not adapt to an emerging technology will face financial losses and bankruptcy. We also saw a similar development regarding the telecommunications corporation Nokia. Nokia was once the leader of the mobile phone market but fell from their pole position after the introduction of smartphone technology. Prior to this development, Nokia's management was well aware of the fact that smartphone technology was affecting the market but assumed erroneously that this would not greatly affect their market share. The smartphones of other companies started to flood the market, increasing their profits and consequently these companies eventually outperformed Nokia. Regarding the introduction of important new technology, there is one strategy that must be adopted, and this is: Adapt, Adapt and Adapt!

What we have tried to point out here is that when creative destruction enters the market, some businesses are destroyed, while completely new businesses based on the new technology emerge and create completely new industries. Sometimes these industries will be an extension of the old industries. At other times, qualitatively new industries will emerge on the basis of the new technology, such as drones, translation software and so on. The workplace of the future will be radically changed not only by new technologies, such as AI applications, but also by the emergence of new skills in relation to the new technological applications. One such example where new skills are needed is that of drone piloting. Drone pilots do not need to have any special insight into artificial intelligence applications. As a general rule, the drones have some form of artificial intelligence built into their design. In other words, the new drone pilots is an example of a job related to the new technology based on artificial intelligence but where the drone pilots do not need to have competence in this technology. Thus, many of the new jobs will not be directly related to AI skills. Against this background, it is no mystery to understand how creative destruction stimulates growth. This growth is a direct result of the old methods that can no longer compete with the new technology, which we mentioned above in the cases of the Swedish company Facit and the Finnish corporation Nokia. The new that is emerging bases its production and distribution on a new technology, new ways of organizing, as well as other innovations that lead to both productivity and earnings being greater than in the old businesses. Investments in the market then move away from the businesses that do not adapt and flow to the new businesses where expectations of greater earnings attract investors.

Based on the narrative, description, analysis, theoretical reflections and the review of the utility value, we have developed a Boudon-Coleman diagram to show some developments related to the workplace of the future (Figure 3.3).

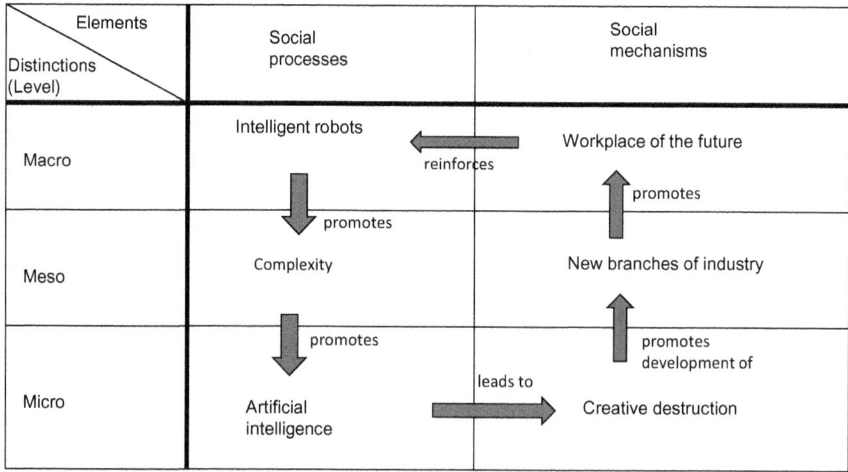

Figure 3.3 The workplace of the future: A Boudon-Coleman diagram.

Systemic connections related to artificial intelligence

Our memory will not be of the same importance in future working life as it has been in the past. Various forms of data storage technology and applications make it possible for us to retrieve very quickly various types of information. For example, we can 'Google', that is, search for information about something on the internet by using the search engine Google.

We will probably be better served by knowing more about connections and relationships between various areas of knowledge than being able to remember factual knowledge. When we know something about connections, we will then be able to ask the right questions, for instance, if we are using a search engine. This will enable us to find the information we need. If this assumption is correct, this will affect how we organize our education system at all levels, from primary to higher education. In other words, memorization will no more be so important for learning, such as rote learning of the names of cities and rivers. As mentioned above, we have technical aids that can do this much better than any human, no matter how good their memory is. This insight tells us something about what future workplaces will demand and what will be less in demand. If this is correct, learning about connections and contexts will be important. At the same time, businesses will be interested in hiring people that understand the importance of knowledge interactions for innovation processes. In other words, innovations emerge when there is an interactive process between different domains of knowledge. This will be especially important in the future because innovation will be the driver of the innovation economy.

Businesses that are unable to generate innovations or to succeed in adapting to innovations generated by others will become uncompetitive. Solving problems has always been important, and will continue to be so in the future. Being able to identify new problems and then finding ways to solve them will probably become more important in the innovation economy than before. The rationale for this claim is that solving problems and generating innovation processes will promote a business' competitiveness in the future innovation economy. In order to generate innovation processes, it will be crucial that employees can identify problem areas that have the potential to create innovations that will be decisive for a business' success. It is in this context that understanding systemic connections can be significant, because it is such connections that can identify crucial border areas between different areas of knowledge that have not previously been in contact with each other. The point of this understanding of connections is based on the insight that new innovations often appear at the boundaries between distinct areas of knowledge. The examples here are many, e.g., information and technology to information technology, biology and technology to biotechnology, communication and technology to communication technology, etc.

Systemic connections also create a side effect of complexity that increases the demand for employees who can simplify this complexity and focus on what the business is designed to do. This increased complexity can reduce the efficiency and productivity of businesses. However, we can reduce this complexity by using AI and robotics. In order to make the best possible use of AI and robotics, it will be advantageous if employees have knowledge of how to develop AI programs that are tailored to work activities and processes. AI and robotics are tools that have a wide range of application. In order to use AI and robotics within more specific areas, professionals need to know how to utilize coding, amongst other things. For example, a psychologist who wishes to develop a therapeutic program for a patient would need to develop specific AI codes that are adapted to the procedures the psychologist is using in his work. This will greatly reduce the complexity of his work while also increasing productivity and efficiency. This insight also implies that the education system that we described above needs to be changed. In such a context, it will be more important to learn how to code and how to use various forms of algorithms, than to memorize various facts, when such knowledge is easily accessible by using search engines.

Human rationality is distinctly different from algorithmic rationality. Human rationality is severely limited while algorithmic rationality is not. When, not if, AI and robotics enter our workplaces, then human rationality will be pushed out and algorithmic rationality will prevail. This can be positive in some contexts but not in all. It will be positive where the decision-making process is related to simple rules that do not involve emotions and social relationships. For example, it can be positive in the case of the buying and selling of shares; calculations in the mass, wind strength and wear and tear of bridge constructions; surgical interventions; prescription processing in pharmacies; self-driving vehicles; drone technology; and so on. Human rationality will be preferable when decision-making is connected to emotional and social intelligence.

Human rationality with its limitations largely corresponds to the complexity of the world around us. In this environment, both emotional and relational factors exist. Reducing these two important aspects of being human could potentially harm both people and social systems. Therefore, it will be advantageous if both the algorithmic rationality and the limited human rationality operate side by side. The algorithmic rationality will promote productivity, while the human rationality will promote the qualitative aspect of future workplaces.

We use artificial intelligence here to refer to algorithmic rationality. It seems reasonable to assume that the workplaces of the future will combine AI applications with the limited human rationality. Even though it may be possible to automate all the human work processes and thereby increase productivity, this will probably not be desirable from a qualitative perspective.

If we have social systems where profit comes before people's well-being, then it is likely that algorithmic rationality will be preferred. In such social systems, human rationality will be downplayed. On the other hand, those social systems that put human well-being before profit will emphasize finding a balance between the two opposing rationalities. In other words, the structure of the social systems of the future will to a great extent depend on political processes and the distinction between profit and human well-being. Political parties can be viewed as being placed along a political spectrum in relation to this distinction, where ends and means will function as a rhetorical superstructure. The various political ideologies will then put forward social strategies based on the distinction and the methods for achieving political goals. This political dimension is important to understand, because our workplaces in the future will not only depend on development of new technology such as AI and automation processes but will also be determined by the political processes and ideologies that govern the development of technology and future workplaces. The workplaces of the future will be designed on the basis of many factors. One factor is the distinction between productivity and human well-being. In addition, there is the distinction between profit for the few or well-being for the many. How these distinctions are structured will determine how wealth creation will be distributed. Therefore, it is not only the development of artificial intelligence and algorithmic rationality that will be decisive for shaping the workplaces of the future, but also which political ideology is adopted.

Regardless of how the balance between algorithmic rationality and human rationality will be achieved, the use of artificial intelligence will become part of most work operations due to global and local competition. As we move towards the Fourth Industrial Revolution, the adoption of AI as the core technology will of necessity lead to creative destruction. The old methods will sooner or later be destroyed, while the new methods will create jobs based on the new technology. These processes will create personal crises for some, while others will be able to come out of this process with a greater degree of prosperity and well-being than before. These processes will develop at different levels and also have an impact on the de-globalization which we see emerging (Johannessen, 2023a).

Based on the above narratives, description, analysis, theoretical reflections, the review of practical utility and the systemic connections, we have developed a loop diagram (Figure 3.4).

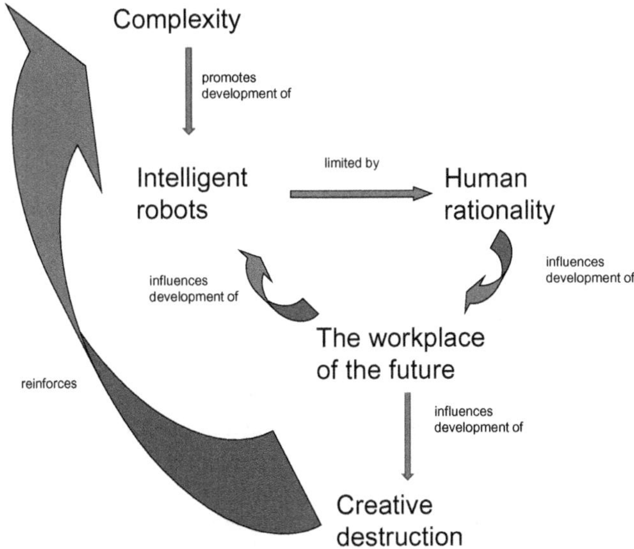

Figure 3.4 The workplace of the future: Some systemic connections - A loop diagram.

Sub-conclusion related to artificial intelligence

In this chapter, we have examined the following question: What is the potential impact of artificial intelligence on future employment trends? We have framed the answer in relation to five elements.

1. Algorithms
   It is very probable that various forms of algorithms, such as learning and evolutionary algorithms, will be more widely used in the workplaces of the future. Such a development means that our education system will need to adapt to the new reality. The abilities to code and to understand algorithmic relationships will be just as important as the three Rs (reading, writing and arithmetic).
2. Nanotechnology
   The intelligent robots of the future will, in most cases, not be humanoid; in fact, most of the AI applications in homes and workplaces will not even be visible to the naked eye. For example, the use of AI could be extended to apparel; AI implemented into clothing would enable smarter clothes more reactive to the body. There will be further development of 'the Internet of Things' (IoT); using sensors connected to the internet, this allows information to be exchanged between people, objects, and devices, such as kitchen appliances. In other words, the IoT will be based on AI and robotics applications which will be invisible to us. The idea of humanoid robots belongs more to the realm of science fiction than the world of science. The intelligent robots of the future will not look like us, but become almost a part of us. For instance, it is most likely the use of different languages will no longer pose a problem. For example, one will have something like an advanced translation program implanted in one's earlobe, so one will be able to receive simultaneous translation of what people who speak different languages are saying and vice versa. The potential areas of application of micro-robotics and nanotechnology are endless. For example, it will be possible to use 'nanobots' to clear clogged arteries.
3. Complexity
   The complexity of our lives will increase greatly. This will occur for various reasons, but possibly the most important explanation is that the world around us is changing rapidly, and more and more systems in the world around us are coming into contact with each other. This complexity is directly linked to our ability to be rational. It has been said by several theorists that human rationality is limited. This can be both positive and negative. The negative aspect is that we fail to be rational when we need to be rational. The positive aspect is that we are not rational where factors other than the purely rational are decisive. These factors can be related to culture, history, emotions, morals, relationships, etc. It is in this context we can say that intelligent robots have an algorithmic rationality and humans have a limited rationality. It is when both types of rationality are used together that we can reduce complexity and make decisions that promote both productivity and human well-being. Rationality

without human well-being is a cold form of existence. Rationality without regard to productivity will hamper prosperity in the short term. It is when both rationalities interact that the creative new can emerge to the benefit of the system as a whole.

4. Creative destruction
   The new technology being developed will certainly lead to old working methods being destroyed and new working methods being created. This creative destruction will be experienced as a crisis by some, while others will use this development to create a better life for themselves. To fully understand how creative destruction affects our workplaces, it may be appropriate to examine the factor of time lag. Time lag relates to the time it takes from when new technology is introduced to when it becomes fully implemented in the economy. It is during this time lag that people and workplaces will need to adapt to the new technology that is emerging. In this context, 'our ship needs to change course' long before it seems strictly necessary. If we wait until the moment when challenges are looming and large, we may find that our social systems are like the Titanic, unable to change course in time. In other words, people, businesses and social systems need to change course and be prepared for the emerging new technology before it seems necessary. This sounds paradoxical, but it is important to be prepared for the moment when creative destruction makes itself felt with full force.
5. Industries of the future
   With the introduction of major new technology, there will still be some old industries that survive, even if they will have to change as the world changes. However, the new technology will also create qualitatively new industries. For instance, during the Second Industrial Revolution, when cars were first used on the streets of New York at the end of the 19th century and the beginning of the 20th century, after a period of time, completely new industries emerged, such as car workshops, gas stations, salvage companies and so on. Similarly, during the Third Industrial Revolution, the new information and communication technologies greatly changed the economy and spawned new industries, such as the production of personal computers, mobile phones, the development of internet and online businesses and so on. When the Fourth Industrial Revolution becomes a reality, completely new industries will also emerge. Of course, we do not know precisely what these new industries will look like around 2030–2050. What we do know, however, is that these industries will be related to AI and robotics technologies and associated technologies.

We can sum up this conclusion by saying that what we know about artificial intelligence and the jobs and workplaces of the future is that they will be dominated by various kinds of robots. These technologies will transform the labour market and the sectors that will emerge on the basis of this technology. We also know that people who succeed in adapting to this emerging technology will be at an advantage in the labour market. Ways of gaining such an advantage could be as simple as learning to code, which could empower us to develop our own small-scale intelligent assistive aids.

Notes

1 https://en.wikipedia.org/wiki/General-purpose_technology
2 Schumpeter, 1951, 1954, 1989, 2002, 2006, 2010.
3 https://en.wikipedia.org/wiki/The_Sciences_of_the_Artificial
4 https://en.wikipedia.org/wiki/Analytics
5 https://en.wikipedia.org/wiki/Brave_New_World
6 Johannessen, 2020, 2020a, 2020b, 2021, 2021a, 2021b, 2022, 2022a, 2022b, 2022c, 2023.
7 https://en.wikipedia.org/wiki/Open_Letter_on_Artificial_Intelligence
8 https://en.wikipedia.org/wiki/Open_Letter_on_Artificial_Intelligence
9 https://www.gartner.com/doc/reprints?id=1-2AOVKCXF&ct=220728&st=sb
10 Kahneman, 2011; Kahneman & Tversky, 1979, 2000.
11 https://appian.com/resources/resource-center/google/gartner-top-strategic-tech-trends-2022-hyperautomation.html?google_ad_keyword=smart%20process%20automation&matchtype=b&google_ad_campaign=1753561383&utm_source=google&utm_medium=cpc&utm_campaign=platform&gclid=CjwKCAjw7eSZBhB8EiwA60kCW4KznOVlY3_HrJA-Z7MsPd3BeyphbBXkXnJngMdvImLcD5ncLY19QxoCQmgQAvD_BwE
12 https://www.mckinsey.com/~/media/mckinsey/featured%20insights/digital%20disruption/harnessing%20automation%20for%20a%20future%20that%20works/mgi-a-future-that-works_full-report.ashx
13 https://assets.kpmg/content/dam/kpmg/us/pdf/2017/10/kpmg-powered-solutions-for-tech-companies-issue-two.pdf
14 Schwab, 2016, 2018; Johannessen, 2020a, 2022a.
15 See prospect theory (Tversky & Kahneman, 1974, 1983, 2000).
16 https://en.wikipedia.org/wiki/Genetic_algorithm
17 https://today.oregonstate.edu/archives/2011/aug/lasting-evolutionary-change-takes-about-one-million-years
18 https://en.wikipedia.org/wiki/Technological_singularity
19 https://english.news.cn/20220526/41f378f2099a4ee7969c81e22046f140/c.html
20 https://english.news.cn/20220526/41f378f2099a4ee7969c81e22046f140/c.html
21 Make a search in the journal *Science Robotics.*
22 https://en.wikipedia.org/wiki/Bounded_rationality
23 https://www.china-briefing.com/news/china-standards-2035-strategy-recent-developments-and-their-implications-foreign-companies/
24 https://en.wikipedia.org/wiki/Made_in_China_2025
25 https://www.voanews.com/a/us-aims-to-hobble-china-s-chip-industry-with-sweeping-new-export-rules/6780765.html
26 https://equitablegrowth.org/a-note-on-niall-ferguson-why-did-keynes-write-in-the-long-run-we-are-all-dead/

References

Adriaenssen, D.J. & Johannessen, J-A. (2016a). Prospect theory as an explanation for resistance to organizational change: Some managerial implications, *Problems and Perspectives in Management*, 14 (2): 84–92.

Aghion, P.; Antonin, C. & Bunel, S. (2021). *The power of creative destruction*, Harvard University Press, Cambridge, Mass.

Delfanti, A. (2021). *The warehouse: Workers and robots at Amazon*, Pluto Press, New York.

Frey, C.B. (2019). *The technology trap: Capital, labor, and power in the age of automation*, Princeton University Press, Princeton.

Harari, Y.N. (2017). *Homo Deus: A brief history of tomorrow*, HarperCollins, New York.
Harari, Y.N. (2018). *21 lessons for the 21st century*, Jonathan Cape, London.
Johannessen, J-A. (2020). *The workplace of the future*, Routledge, London.
Johannessen, J-A. (2020a). *Automation, innovation and economic crises: Survival the fourth industrial revolution*, Routledge, London.
Johannessen, J-A. (2020b). *Artificial intelligence, automation and the future of competence at work*, Routledge, London.
Johannessen, J-A. (2021). *China's innovation economy: Artificial Intelligence and the New Silk Road*, Routledge, London.
Johannessen, J-A. (2021a). *Artificial intelligence, automation and ethics in the innovation economy*, Routledge, London.
Johannessen, J-A. (2021b). *Communication as social theory: The social side of knowledge management*, Emerald, London.
Johannessen, J-A. (2022). *The new silk road and the innovation economy in China*, Routledge, London.
Johannessen, J-A. (2022a). *Creativity, innovation and the Fourth Industrial Revolution: The da Vinci strategy*, Routledge, London.
Johannessen, J-A. (2022b). *A systemic approach to continuous change in the innovation economy*, Routledge, London.
Johannessen,J-A. (2022c). *Intelligent robots consciousness and creativity: The search for hidden knowledge, the cognitive side of knowledge management*, Emerald, London.
Johannessen, J-A. (2022d). *The philosophy of tacit knowledge*, Emerald, London.
Johannessen, J-A. (2023). *Feudal capitalism in the innovation economy*, Routledge, London.
Johannessen, J-A. (2023a). *De-Globalization in the innovation economy: Technological innovations trigger conflict between China and the United States*, Routledge, London.
Johannessen, J-A. (2024). *Safe jobs in the innovation economy: Strategic competence creation*, Routledge, London.
Jones, P. (2021). *Work without the worker: Labour in the age of platform capitalism*, Verso, London.
Kahneman, D. (2011). *Thinking fast and slow*, Allen Lane, New York.
Kahneman, D. & Tversky, A. (1979). An analysis of decision under risk. *Econometrica, Journal of the Econometric Society*, 47 (2): 263–292.
Kahneman, D. & Tversky, A. (2000). Prospect theory: An analysis of decision under risk, in Kahneman, D. & Tversky, A. (eds.), *Choices, values and frames*, Cambridge University Press, Cambridge, pp. 17–43.
Kalia, P. (2016). Artificial intelligence in E-commerce: A business process analysis, in Bhargava, C. & Kumar Sharma, P. (eds.), *Artificial intelligence. Fundamentals and applications*, Routledge, London, pp. 9–19.
Kotkin, J. (2020). *The coming of neo-feudalism: A warning to the global middle class*, Encounter Books, New York.
Kumar Singla, Aggarwal, V. & Gupta, S. (2021). Artificial intelligence and nanotechnology: A super convergence, in Bhargava, C. & Kumar Sharma, P. (eds.), *Artificial intelligence: Fundamentals and applications*, Routledge, London, pp. 1–9.
Kurzweil, R. (2005). *The singularity is near*, Penguin, London.
Kurzweil, R. (2008). *The age of spiritual machines: When computers exceed human intelligence*, Penguin, London.
Kurzweil, R. (2013). *How to create a mind: The secret of human thought revealed*, Penguin Books, New York.
McDonagh, S.C. (2021). *Robots, ethics and the future of jobs*, Messenger Publication, New York.
Minda, J.P. (2020). *The psychology of thinking*, Sage, London.
Mishra, M.K.; Mishra, B.S. Patel, Y.S. & Mishra, R. (2019). *Smart techniques for a smarter planet: Towards smarter algorithms*, Springer, London.

Nietzsche, F. (1968). *Twilight of idols*, Penguin, London
Ross, A. (2016). *The industries of the future*, Simon & Schuster, New York.
Schumpeter, J. (1951). *Theory of economic development*, Harvard University Press, Boston.
Schumpeter, J. (1954). *History of economic analysis*, Oxford University Press, Oxford.
Schumpeter, J. (1989). *Business cycles*, Porcupine Press, New York.
Schumpeter, J. (2002). Seventh chapter of the theory of economic development, translated by U. Backhaus, *Industry and Innovation*, 9: 93–145.
Schumpeter, J. (2006). *Theorie der Wirtschaftlichen Entwicklung*, Dunker & Humblot, Berlin.
Schumpeter, J. (2010). *Capitalism, socialism and democracy*, Routledge, London.
Schwab, K. (2016). *The fourth industrial revolution*, World Economic Forum, Geneva.
Schwab, K. (2018). *Shaping the fourth industrial revolution*, World Economic Forum, Geneva.
Simon, H.A. (2019). *The sciences of the artificial: Reissue of the third edition with a new introduction by John Laird*, MIT Press, Boston.
Sønderholm, J.; Mainz; J. & Uhrenfeldt, R. (2021). Big data analytics and how to buy an election, *Public Affairs Quarterly*, 35 (2): 119–139.
Tversky, A., & Kahneman, D. (1974). Judgment under uncertainty: Heuristics and biases. *Science*, 185, 1124–1131.
Tversky, A., & Kahneman, D. (1983). Extensional versus intuitive reasoning. The conjunction fallacy in probability judgment. *Psychological Review*, 90, 293–315.
Zhou, Z-H-; Yu, Y. & Qian, C. (2019). *Evolutionary learning: Advances in theories and algorithms*, Springer, London.
Zuboff, S. (2019). *The age of surveillance capitalism*, Profile Books, London.

4 Innovation

Core idea in this chapter

As a general rule, it seems to be co-creativity between innovators, entrepreneurs and public institutions that enables the establishment of an institutional framework that fosters innovation and economic growth.

Key points in this chapter

- The information available to us steers our thinking and our actions.
- The power dynamics of the future will be determined by control over Big Data.
- Economic inequality will boost the trend for increasing focus on consumption.
- Big Data, artificial intelligence and analyses performed by intelligent robots will soon be capable of predicting our behaviour in various situations, long before we ourselves know how we will act.
- In the Fourth Industrial Revolution, the abilities to code and to understand algorithmic relationships will be just as important as the three Rs (reading, writing and arithmetic) were to those who experienced the First, Second and – to some extent – Third Industrial Revolutions.
- The ability to edit human DNA may be the most important medical innovation of the 21st century. On the other hand, this innovation could also bring about the most serious challenges for the whole of humanity.
- If people believe that feudal capitalism is in their best interests, we will see the emergence of a completely new democratic phenomenon: A feudal capitalist bureaucratic machine.
- Robot shock transforms people into quantities of data within huge databases. In this way, their individual actions have no autonomous value, because they are structured into pre-designed models that determine how they will act.
- Intelligent robots will impose significant restrictions on our freedom to make our own decisions.
- There is a qualitative difference between having a model of something and developing a model in order to achieve something.

DOI: 10.4324/9781003567325-4

Introduction

In this chapter, we will examine the following question: What is the potential impact of innovations on future employment trends? The Fourth Industrial Revolution has the potential to transform workplaces, democracy and other institutions. Robot shock that relates to innovations of which we can already see the outlines will have an impact on our understanding of the nature of democracy and freedom. In election after election in both the US and Europe, we see how democracy and freedom are directly related to our decision-making options at an individual level.[1] Our political perspectives will be determined largely by our perception of what gives us these decision-making options. The danger here is that new technology linked to Big Data and the ways this data can be used gives some people the power to control people's perceptions more precisely than was previously the case. If it is correct that it is the information we have available that guides our thinking and our actions, then this new technology could determine our behaviour in elections and other situations. It could even be the case that we will argue in favour of economic inequality, even though this is against our own interests, because the information available to us indicates that inequality is the best outcome for all of us.

The basic assumption behind the argument above can be expressed in the proposition: We create our own reality with what we think, based on the information and knowledge that is accessible to us. This proposition can be related to the classic proposition that knowledge transforms the knower. Implied by these two propositions is that the freedom we think we have about how we choose to view the world is actually controlled by those who have control over large amounts of data (Big Data). It is in this context that the robot shock can really affect our everyday life. We think we make free choices, while we are actually governed by various technological innovations, which control all available data (Carou et al., 2022; Reich, 2009). It is precisely the access to Big Data that will determine the power dynamics of the future. As we have mentioned before, but it cannot be emphasized enough: In the agricultural society, it was the ownership of land that determined power relations. In the industrial society, it was the ownership of capital and capital goods that determined power relations. In the innovation economy, as we move towards the Fourth Industrial Revolution, it will be knowledge and data that will be the central factors determining power relations.

When innovations, both technological and institutional, enter our workplaces, it will necessarily lead to some succeeding in the competition, while others fail. The basis for some surviving while others fail to adapt to the new innovations is the competition that exists in the market. When businesses are unable to adapt and go bankrupt, this naturally leads to changes for the workers and the businesses. These changes can lead to minor and major financial crises. These crises, as well as the fact that people's skills need to be rebuilt, will lead to an increase in economic inequality. This economic inequality is largely based on the fact that the skills of many people are no longer competitive in relation to the new technology. Those who can quickly update their skills so they are compatible with the new technology and the new innovations in the market will be able to adapt to the new times.

An important point in this context, however, is that the innovations that change the market also increase productivity in our workplaces. The technology automates many work processes and thus creates less need for labour in these work operations. On the other hand, it seems that the demand for labour will rise, because there are many work operations that these machines are unable to perform. These work operations, on the other hand, will be in professions and occupations that do not have such high productivity. We know from economic history that productivity has always corresponded to higher wages. If it is true that the new innovations promote productivity in some industries but not in others, then this will also mean that economic inequality will be a consequence of the 'robot shock'.

'Robot shock' will not necessarily be experienced as something negative by most people. In other words, we are not saying that the robot shock will lead to a future dystopia, the worst of all worlds. Nor are we saying that the robot shock will lead to a wonderful utopia, where everyone is well off. The robot shock has in it the potential that many people will be able to improve their lives, but there is also the possibility that many people will be worse off. The robot shock will not be a deterministic event. It will constitute a scenario where we can shape the future ourselves. Unfortunately, however, if we consider the propositions above, related to how available information and knowledge can influence our thoughts and actions, and that knowledge can change us, then it may appear that our free choices to influence our own future will be limited, and we will not be as free as we would like to believe.

A clear danger with innovations, especially technological innovations and institutional innovations, is that people become passive and end up as submissive consumers. The reason is that there will be a greater focus on income as the new technologies promote automation processes. In other words, economic inequality will boost the trend for increasing focus on consumption.

As automation increases and becomes a significant part of our workplaces, both work processes and consumption will become standardized. It is standardized operations and standardized products that will result in those people at the bottom of the income hierarchy having to buy cheap standardized products. These products will mainly be produced in countries where costs are low. This also implies that the standardization will promote more standardization, and those people who demand this will in fact just be making things worse for themselves, because both the jobs and skills will be drawn towards the countries where production takes place. This does not apply to everyone, but to those who are lowest on the wages hierarchy. The paradox is that when people demand standardized products, which are of course much cheaper than more tailored products, from an overall social viewpoint they are undermining their own economic position. This may be likened to sawing off the branch you are sitting on.

We already see today that it is not just standardized products that we buy from countries such as China. Our entire culture is being standardized. Everyone is streaming the same movies. Although the movies may have different content, they convey much of the same cultural values. It is the information we get from the distinct but linked cultural expressions, which creates our understanding of the world around us. It is against this background that we also act in relation to the outside world. We are transformed through the input of information and create the knowledge that sustains our perceptions. If this argument proves to hold true, then our choices will be determined by information that others create.

When Big Data, artificial intelligence and analyses performed by intelligent robots are capable of predicting our behaviour in various situations long before we ourselves know how we will act, then our freedom of choice will be limited. The paradox is that we believe we make free and informed choices. This social and psychological pressure does something to the individual. We believe we are free actors who make choices based on a rational decision-making process. The reality is that our freedom to make choices is being constantly restricted. We can go all the way back to Russel Ackoff, who, in 1981, wrote the book: *Creating the Corporate Future: Planned or be Planned for*. The point in this context is that being able to plan one's own future or the future of a business has been severely restricted by the development of the new technology. In our context, this means that innovations affect how we think and act. When access to data becomes the new power, you can say that this data becomes information through various processes. It is then this information which determines how you act. If this information is monopolized, then it is also highly probable that those people who own the data that becomes information will be able to control our behaviour. Thus, we will have reached the unavoidable situation – we can no longer plan our future as this will be 'planned' by others. It is those people who own data about us and about our choices of action that will plan our future. Our freedom of choice in such a context will be no more than mere rhetoric, which other people propagate to make us believe that we are 'free', and that we imagine we have our own thoughts and can decide what our actions will be. This idea has been expressed by the philosopher Nietzsche (1968: 35) in his book *Twilight of the Idols* in which he greatly criticized the culture of the day as corrupt and destructive.

Narratives related to innovation

Case letter 1: The 'conceptual gold standard': An index for the development of the workplace of the future.

Innovation, automation, productivity and employment are key concepts when discussing the workplace of the future. These concepts are so decisive for how our future workplace will develop that we term them the 'conceptual gold standard' when investigating the workplace of the future.

This gold standard can be linked to the robotization of our workplaces. It is important to understand that the intelligent robots of the future and the present have nothing to do with science fiction literature's humanoid robots. The intelligent robots that will take over most, if not all, work functions will be nanorobots, and the like, that are not visible to the naked eye. These intelligent robots of the future will be used in our everyday devices and applications, such as in our clothing and smartphones. The robots are already used in our production processes and implemented in various forms of transport, such as boats, cars, planes, drones and so on. There will be a further development of 'the Internet of Things' (IoT) using AI and sensors connected to the internet, allowing information to be exchanged between people, objects and devices such as kitchen appliances.

One can go so far as to say that the conceptual gold standard for the workplace of the future can be designed as an index that says something about the development of the specific workplace. This index could be developed by examining the degree of innovation, automation, productivity and employment development at the specific workplace. Employment development is probably the most difficult component of this index. One can conceive, however, that the employment conditions will be determined by the degree of competence that the employees possess. In this way, one could say that it is the level of competence of the employees that says something about the quality and compatibility of the employment. We are aware that this is not necessarily true for all workplaces in the future, such as many jobs within the service sector. However, the degree of competence level will be able to tell us something about how the specific workplace will be able to develop in the future.

The new innovations in artificial intelligence, Big Data, learning and evolutionary algorithms and so on, have created the basis for a qualitatively new era of automation. However, we would like to emphasize that this new wave of automation will not threaten employment to the same extent as previous waves of automation. As we have shown in the previous two chapters, this will lead to the destruction of some jobs, but it will also, as has been shown throughout economic history, create more jobs that are compatible with the new technology. However, what seems to be happening, and is already a fact, is that income inequality will increase. Those people with a high level of competence and an education that is compatible with the new technology will be more highly paid. Those people with little education will end up at the bottom of the income scale. On the other hand, there are many indications that there will always be a need for hands-on staff that can take care of clients and users, such as healthcare workers in daycare facilities, nursing homes and other institutions.

It may prove difficult to use intelligent robots in work processes that are related to human emotional and social care. Attempts have already been made to overcome this emotional and social barrier by developing robots that can communicate with elderly people in need of help in Japanese care centres. So far, however, these experiments have produced varied results and not been wholly satisfactory.

We know that automation processes have been under development for several hundred years; at least since the start of the First Industrial Revolution (1750). The new development that is occurring, as we move towards the Fourth Industrial Revolution, is that the rate and direction of automation has increased.

We are aware of the fact that automation and productivity may raise more questions than anyone is able to answer. What we do know, however, is that technological innovations affect automation processes and promote productivity in many workplaces. At first glance, this may seem intimidating to those who do not have the necessary and compatible skills.

We have largely thought of humans as possessing the most advanced cognitive capabilities and skills. This is probably a perception we will have to reformulate, when intelligent robots can acquire cognitive skills of algorithmic rationality far better than any human can with their limited rationality. The point, however, is that our limited rationality applies precisely within the cognitive area, not the emotional or social areas. Thus, our limited rationality becomes an unlimited advantage for us humans in an age where artificial intelligence, machine learning, learning and evolutionary algorithms far surpass us in cognitive rationality. What could be perceived as a limitation in our rationality seems to become a strength for people at a time where closeness, care and social capabilities are becoming increasingly important.

For the individual business, automation will very likely promote the performance of the business as a whole. For the individual in this workplace, however, things appear somewhat different. There will be a requirement for a changed competence profile in the businesses that automate and increase efficiency. For the individual, it can quickly become a question of skills development or finding something else to do. The more adaptable the individual is, the greater the probability that he/she will only remain unemployed for a brief period.

Another aspect of AI technology that leads to increased productivity is the improvement in product quality. We can suggest the following proposition: The more AI applications an organization uses, the greater the probability of higher quality products. This proposition is related to the fact that there will be fewer errors in the production process.

Most economists and economic historians would agree that throughout history an increase in productivity has led to economic growth and higher wages. The future will show whether the increase in value creation will result in wage growth for most people or whether it will be mainly distributed to capital owners and those people with compatible skills. The distribution of value creation will thus become a political issue. However, it is highly probable that increased automation will increase global productivity by 0.8–1.4 per cent annually.[2]

Description related to innovation

In the innovation economy, as we move towards the Fourth Industrial Revolution, it seems that a central proposition is valid: The more innovations that enter the market, the fewer choices people will have. This is a counterintuitive proposition, and thus needs a little explanation. We are used to thinking that innovation promotes our choices, because there is a greater range of possibilities to choose from. However, as suggested by the above proposition, when major technological innovations enter the market, we are all compelled sooner or later to become consumers of these innovations. A good example is mobile phones. The continuous development of smartphones enabled them to fulfil many functions, so that people became locked into a consumption pattern where ‘everyone had to have a smartphone in order to function in everyday life. However, people felt that their options had increased, and they believed themselves to have ‘free choice’ because there were several brands and models to choose from. Despite this idea concerning free choice, everyone became dependent on having a mobile phone of one type or another. In reality, despite the idea of free choice, the phones are greatly similar to one another and fulfil the same functions more or less. People feel they have greater choice, but this is within the same product area. With this understanding of innovation, we can argue that the technological innovations that will dominate the market as we move towards the Fourth Industrial Revolution will limit, not expand, our choices. This line of argument agrees with some of the ideas of the French philosopher and sociologist Jacques Ellul, which he expressed in his book, *The Technological Society* (1967), where he shows how technology has changed from being the servant of humanity to becoming something that threatens humanity – a technological tyranny.[3] 'Robot shock' viewed in this light, that is, the development towards a technological tyranny over our thoughts and actions is not an uplifting scenario. In such a context, one can conceive that people will become consumers in a feudal state, where technological and social domination becomes possible.

Innovation can be understood from different aspects. The aspects we choose to focus on are related to five elements. These five elements, which we will describe further in the following, are: economic and social crises, feudal structures, career opportunities, creative processes and 'hacking humans.

Economic and social crises

There are many ways to understand the process of value creation. Innovations, which drive value creation, are one way of understanding how value is created in a society (Baumol, 2004). It is precisely innovation that is the main driver of value creation in the innovation economy.[4] It has become evident to most economists and social scientists that in order to achieve economic growth based on innovation, it is essential that there is an interaction between private entrepreneurs and public institutions. The innovation economy based on this interaction will transform society in completely different ways than has been the case previously. Mariana Mazzucato is one of the central theorists who has examined this problem area in recent years (2013, 2019, 2021). When economic growth driven by revolutionary innovations transforms the economy, people and businesses are subjected to economic and social crises. This occurred during the first, Second and Third Industrial Revolutions, which were all driven by major technological innovations. The First Industrial Revolution, which began around 1750, was driven by the steam engine; the Second Industrial Revolution, which began around 1870, was driven by the internal combustion engine, electricity and mass production; and the Third Industrial Revolution, which began around the 1950s, was driven by the rise of digital technology and the internet. The Fourth Industrial Revolution will be characterized by robotics, artificial intelligence and related technologies.[5] It is interesting to note that the various industrial revolutions occur every hundred years, more or less. Kurzweil (2005, 2008, 2013) suggests that the technological singularity (Shanahan, 2015) will occur sometime around 2035–2045. Technological singularity here refers to when intelligent machines become more intelligent than humans (Johannessen & Sætersdal, 2020).

It is these industrial revolutions that provide the empirical background for our statement that major innovations lead to major economic and social crises. The introduction of major new technology results in the redundancy of 'old' skills and methods and the transformation of workplaces. Moreover, the introduction of minor innovations can lead to minor crises. Economic history shows many such examples. Previously in this book we have mentioned the Swedish company Facit. We have also described Nokia and the case regarding Kodak and Instagram. However, in these cases, the innovations that resulted in a loss of profits and bankruptcies cannot be characterized as revolutionary. They were largely minor innovations, which led to minor transformations. These transformations created a demand for new skills and new ways of organizing. The transformations also made it clear that in order to survive in the emerging innovation economy, an integration of private entrepreneurs and public institutions was important in order to survive the crises.

The workplaces of the future will be greatly impacted and transformed by AI and related technologies. Of course, we do not know exactly what this transformation will look like in detail. What we do know, however, is that there will be a connection between innovation, automation, productivity, employment and wage development. It is the link between these five elements that will determine the development of future workplaces. What we also know is that the crises will negatively impact those people and businesses that are unable to develop skills and expertise that are compatible with the new technology. This is one of the reasons that the abilities to code and to understand algorithmic relationships will be just as important as the three Rs (reading, writing and arithmetic) were to those who experienced the First, Second and – to some extent – Third Industrial Revolutions. It is on this basis that a strong cooperation between private entrepreneurs and public institutions will be crucial in order to cope with developments in the Fourth Industrial Revolution.

Feudal structures

When social structures become more frozen and social mobility between the various social strata is reduced, it is not improbable that social development will also be hindered. On the one hand, new technology can be developed to carry out various work functions. On the other hand, this will result in the redundancy of those workers who previously performed these work functions.

We have seen a development in the US where society has come under severe pressure as a result of global economic development. This has led to feelings of frustration, powerlessness and alienation spreading throughout the population (Frank, 2004, 2020; Carney, 2020). At the same time, there has been a decline in the American middle class (Reeves, 2018). A growing underclass and greater economic inequality have also characterized American development in recent years, leading to increased domestic tensions.[6] In other words, this social development began long before the implementation of new automation technologies, such as AI and robotics. Kotkin has termed this development of new social structures 'neo-feudalism' (Kotkin, 2020). In this book, we choose to use the term feudal capitalism (Johannessen, 2023). These new feudal structures are emerging not only in the US but also in several European countries.

Of course, it is not the case that history is repeating itself so we are returning to the feudal society of the Middle Ages. However, the new emerging social structures and the lack of social mobility can be compared to the social structures of the feudal society. Nevertheless, it should be pointed out that capitalism as an economic system emerged some centuries after the end of the feudal age. Nonetheless, the paradox that seems to be emerging is that capitalism is taking on a new feudal form. There have been various types of capitalism in the past. The first period of capitalism is termed mercantilism or trade capitalism (broadly from 1500 to 1750). Industrial capitalism emerged around about 1750. Finance capitalism is a stage of capitalism in which economic domination is exercised by financial institutions rather than by industrial capitalists. Finance capitalism is also characterized by the dominant role of large corporations. This development meant that the way towards more feudal structures was a short one. Feudal capitalism tends to freeze social structures and reduce social mobility. It is precisely social mobility that has opened up opportunities for education to function as a social mechanism to enable people to become part of the middle class. When the middle class is reduced, such as in the US (Reeves, 2018), then the opportunities for a 'class journey' are also reduced. This development can be interpreted from research conducted by Acemoglu and Autor (2011) and Ainley (2014, 2016). In the past, education represented a path towards a secure job. Today, higher education is no longer a guarantee of a future position in the middle class. On the other hand, there is much to suggest that many people with higher education will end up in what Standing (2014, 2014a, 2016, 2019) terms the precariat. In other words, higher education no longer provides a secure way into the middle class. It is in this context that Ainley (2016) says that we are betraying an entire generation, because higher education is failing young people and no longer provides graduates with a secure well-paid job.

Figuratively, we can imagine that the social strata resemble great bulkheads that are separated by iron doors closing slowly but surely. Most people will remain in the lower bulkheads in this picture. There will always be some who manage to move between the fixed bulkheads. This is also related to the idea that everyone has the opportunity to move up in the social hierarchy if they work hard enough. However, this no longer holds true, because no matter how much most people try, only a few will manage to move socially upwards. Moreover, in the future, the majority will encounter a workplace that has become more automated and where the key to getting a secure job will be a long higher education that is compatible with the new technology.

Thus, the feudal structures have slowly developed from finance capitalism to what now appears as a new form of capitalism, feudal capitalism (Johannessen, 2023). The ideas for this capitalism are implicit in the capitalist project. This can be simply described as: Profit before welfare, centralization of capital and fewer people making the large strategic decisions.

Career opportunities

It is nothing new that people fear losing their jobs. Throughout history, technological and other types of innovation have nurtured this fear. This has been the case since the first innovations created the capitalist economy in the 17th century in the Netherlands. Dutch organizational and institutional innovations changed the conditions for investment in the trade with East Asia. Amongst these innovations was the foundation of the Amsterdam Stock Exchange and the establishing of joint-stock companies. The first market bubble was the Tulip Mania Crisis. When these innovations spread to England and France, other financial bubbles also developed, such as the South Sea Bubble and the Mississippi Bubble (Johannessen, 2018). When the First Industrial Revolution saw the light of day around 1750 in Britain, technological innovations were introduced that radically affected workplaces (Allen, 2009; Mantoux, 1961). It is therefore nothing new that innovations impact workplaces. What is new in the innovation economy is that a qualitatively new technology, which we can term a general purpose technology (GPT), is driving changes.[7] General-purpose technologies (GPTs) are technologies that can impact an entire economy, both nationally and globally. The archetypal examples of GPTs are the steam engine, electricity, the internet, rail transport, information technology and the computer. Economic history and economic theory shows us that GPT technologies first reduce productivity, before productivity increases sharply. Robert Solow, who was awarded the Economic Sciences Nobel Prize, observed that when more investment was made in information technology, productivity was initially reduced (the productivity paradox).[8] Regarding the digitization of the economy, many commentators question if this 'productivity paradox' is appearing again.[9] There are many explanations of the productivity paradox. For example, it takes time before people learn new skills so that the technological innovation can be as effective as it has the potential to be. Another explanation is that transformation processes that take place in organizations are not necessarily rational when old skills and methods are being replaced by new skills and methods. It also takes a long time before this chaotic organizational process is completed, and a new organizational process has become normalized.

Many commentators claim that artificial intelligence (AI) will become the next general purpose technology (GPT). This technology has the potential to change not only the global economy but also the geopolitical situation. This is evident in the development and production of microchips, which are crucial for the various applications of AI technology. These chips are largely produced in South Korea and Taiwan. This can partly explain the rising tensions today between the US and China, regarding the relations with Taiwan. Taiwan (the Republic of China) claims it is a de jure sovereign state, while China claims that 'reunification' must be fulfilled and that Taiwan has been part of China since the Chinese Communist Party came to power in 1949.

Moreover, today's technological innovations have the potential to influence the development of the workplace of the future. The development of artificial intelligence is also creating tensions in the global economy, which is triggering tensions between the two largest economies in the world, the US and China. Regarding career opportunities, it seems reasonable to assume that those job seekers with extensive higher education that is compatible with the new technology can look forward to a bright future. The robot shock will result in this group of job seekers being able to find permanent, secure and well-paid jobs in the future. However, those without such qualifications and who do not have skills that are compatible with the new technology will face severe challenges in the future labour market. In his book, *A World Without Work* (2020), Daniel Susskind writes about future technology and automation and how we should respond when new AI technology transforms the world of work. However, contrary to Susskind, we do not believe that artificial intelligence will lead to mass unemployment. On the other hand, we can agree with Robert Solow's 'productivity paradox', and when he said that 'You can see the computer age everywhere but in the productivity statistics'.[10] On the basis of this understanding, we have formulated a proposition associated with Solow's ideas:

Proposition: The effects of automation resulting from the deployment of artificial intelligence is not visible in the unemployment statistics but in the wage statistics.

Creative processes

We know little about the domain of knowledge that lies behind the development of creativity and innovation. What we do know, however, is that creative processes are a necessary prerequisite for creating the new, that which the world has never seen before. Some of these inventions become innovations but far from all. It is the application in a market and the market's acceptance of the invention that makes it into an innovation. This argument is fully in line with the definition of innovation by Zaltman et al. (1973): 'any idea, practice, or material artefact perceived to be new by the relevant unit of adoption'.

Ideas are seen as the smallest unit in the innovation process (Hamel, 2002, 2012). However, this refers to the ideas that are in process of development and not fully developed ideas. Before an idea can be characterized as innovative, it must prove to be beneficial to somebody, i.e., the market must accept the idea and apply it. Consequently, the creative process of innovation is here understood as the benefit it has for a market (Amabile, 1996). Thus, it is not sufficient that an idea is new for it to be considered an innovation. An idea may have a great degree of novelty, but if it is of no benefit to anybody in the market, then it has no innovative value (Johannessen et al., 2001).

It is a recurring debate whether it is the private sector or the public sector which is the most important driver of innovation that leads to economic growth. In her book, *The Entrepreneurial State* (2019), Marian Mazzucato stresses the importance of the state's role in the development of innovations. According to conventional wisdom, innovation processes are best fostered by the entrepreneurs in the private sector and do not need government interference. In other words, it is the entrepreneurs who create innovations and value creation, while it is the public sector that uses this value creation for the benefit of the population. Thus, the argument presented is that without private investment and profits from innovation, there would be no material resources that would be available for the public good. Mazzucato argues against this belief saying that this is just a myth constructed by the private sector in order to make financial gain. It is the state, through its education system, funding, vision development and relations with other countries, which largely sets the conditions for innovation processes in the private sector. Mazzucato argues that the development of expertise is one of the most important factors in the development of innovation. This argument seems even more relevant as we move towards the Fourth Industrial Revolution. The rationale is that if innovations are important for value creation, then expertise related to the new technology will be a decisive driver in the innovation economy. Artificial intelligence requires many different types of skills and expertise in order to develop and apply this technology.

Mazzucato (2019: xxi) claims that public investment in education, research and technological development has been a fundamental prerequisite for economic progress. This argument applies to both the West and in China. In China, this is explicitly expressed in the large investments made in mass education and competence building at Chinese universities from the 1980s up until today.[11] Mazzucato argues that the public sector in the West (US and Europe) has played an almost equally decisive role as it has in China in the development of innovation and economic growth. She argues that this applies to three areas, in particular, the internet, the health sector and the green economy. It is public funding that has laid the foundation for a long-term strategy that has made innovations possible within these three areas, argues Mazzucato. She adds that the risk-averse private sector often makes investments after the state has made the high-risk investments. It can also be argued that there have been similar developments in the Chinese economy from the 1980s up until today (Johannessen, 2021, 2023a).

In this section, we have attempted to stress that creative processes are crucial for the development of innovations. However, these creative processes depend on research projects that are often funded by public investment. Once research and expertise has been developed, it can then be utilized in both private and public projects. In the innovation economy, as we move towards the Fourth Industrial Revolution, artificial intelligence will constitute the key technology. This technology is so complex and encompasses such different areas of expertise and research, with a high risk of failure, that the public sector will have an increasing role in developing the necessary expertise to develop this new technology.

'Hacking humans

The phrase 'hacking humans' is taken from Harari's (2022: vii) introduction to a new edition of Huxley's *Brave New World*.[12] The phrase relates to the technological possibility of editing human genes (Davies, 2021). One can only imagine how this technology will be used in the future to develop various types of innovations to serve someone's special interests.

Basically, being able to edit human DNA can be an important medical innovation. Gene therapy can be used to cure or treat diseases (Doudna & Sternberg, 2018). CRISPR (Clustered Regularly Interspaced Short Palindromic Repeats) can be used to cut DNA and thus be used as a gene-editing tool by editing genes out of the gene pool (Davies, 2021: 3–17). The negative aspect of this technology is very easy to point out. It is easy to imagine how authoritarian regimes might be interested in editing the genome of certain people for certain purposes, such as soldiers. One can also imagine that large companies may be interested in very special types of workers in the working life of the future.

Harari (2022: vii) writes that there are three areas of expertise that are crucial to editing genomes. These three areas are:

1. Biological competence
2. Large amounts of data (Big Data)
3. Large computing power

It is mainly governments and large enterprises that are able to develop these three areas of expertise. The three areas can be advantageously linked to a fourth area: Artificial intelligence. When AI has been further developed, gene-editing will become an applied science that governments will be able to utilize for various reasons, such as altering the gene structure of certain categories of people with specific functions. It is with such a future scenario we can begin to envisage the contours of Huxley's dystopia, *Brave New World.* In Huxley's novel, people are biologically engineered into specific social hierarchies in order to serve the interests of the government. In this context, one can perhaps refer to another dystopia that warned of dismal future developments: George Orwell's (2008) *Nineteen Eighty-Four.*[13] Orwell's novel describes how new technology may be used in the future by dictatorial regimes to oppress the populace by monitoring and controlling citizens. In the present day, key figures, such as Stephen Hawking, and other scientists, have signed an open letter describing the possible dangers associated with the unfettered future development of AI technology.

When artificial intelligence, biological competence, Big Data and very powerful computers are connected together to edit human genes, we will have a technology that has the potential to completely change social systems. If we also relate this development to another development, which we have described and analysed, namely the rise of feudal capitalism (Johannessen, 2023), then we can really begin to envisage the contours of the engineered social structures, which Huxley wrote about. If we also relate this technology to the control systems described by George Orwell in his novel, then we have really come very close to a feudal capitalist society, which may very well characterize the new millennium (Johannessen, 2023, 2024a).

It can be imagined that if such technology existed in Nazi Germany or Stalin's Soviet Union, much would have been different today. We can also imagine how the present-day authoritarian governments, such as Russia and China, may use this

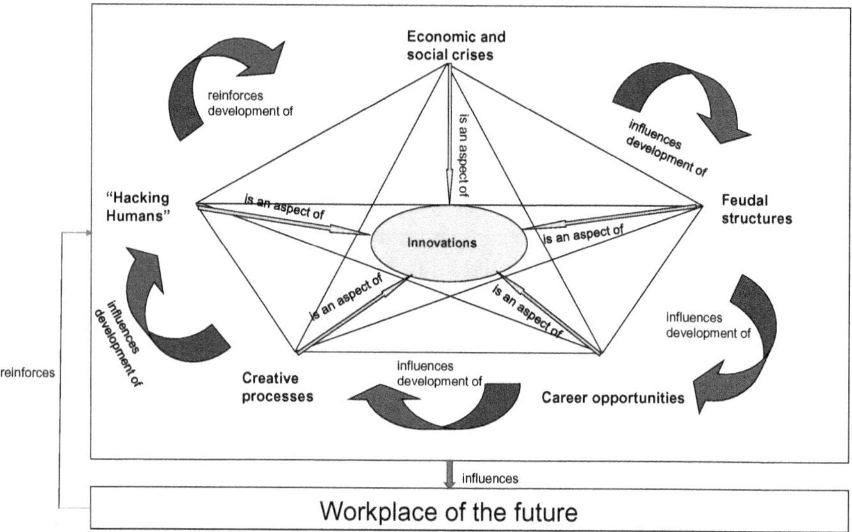

Figure 4.1 Aspects of innovation that will affect the workplace of the future: A conceptual model.

technology in order to promote their own interests. The point of referring to feudal capitalism is that it is not only authoritarian governments that will be able to use this technology to promote their own interests. If feudal capitalism and feudal capitalist ideology become part of the Western way of thinking, then there will also be so-called democratic countries that use this technology to develop various types of innovations. It can also be imagined that social hierarchies will emerge resembling watertight social bulkheads and the creation of gene-edited 'guardians' who will function as a new type of 'gatekeeper' between the closed social bulkheads.

We are more fearful of an ideology where most people believe it is expedient to develop a form of democratic feudal capitalism, where people sincerely believe that these feudal structures are to their own advantage, than of a future Stalin, which Harari (2022: viii) points out. If people believe that feudal capitalism is in their best interests, we will see the emergence of a completely new democratic phenomenon: A feudal capitalist bureaucratic machine.

Based on the above description, we have developed a conceptual model, which combines the five elements that we have discussed (Figure 4.1).

Analysis related to innovation

The use of algorithms in the execution of work processes has increased over several years.[14] The examples are many: Banking, finance and manufacturing are some of the sectors where algorithms control many of the decisions related to work processes. These algorithms are technical innovations that have directly and indirectly affected people's working lives in various industries. As artificial intelligence and intelligent robots occupy more and more areas of working life, these technologies will affect our workplaces with increasing force. Facebook, Google, Netflix and other technology companies use algorithms to get closer to their customers.

Algorithms identify information about our purchasing behaviour. In other words, they are used to construct models of our previous purchasing behaviour. If we assume that intelligent algorithms are developed that can design models not only of our past purchasing behaviour, but models that can predict our future purchasing behaviour, then these algorithms will be able to function as a type of 'Oracle of Delphi' for the big tech giants. They have data about us which they can use in a larger computer system so that data about people become integrated in Big Data related to the purchasing behaviour of billions of people in different situations. Using a Big Data system, the large technology companies can obtain the information they need about people's future purchasing actions. In other words, Big Data in conjunction with intelligent algorithms and robotics can be used to predict consumer behaviour.

Adam Smith's book, *The Wealth of Nations* (1776), is viewed by many as providing the foundation for modern economics.[15] The book emphasized the importance of people's self-interest for economic growth. In other words, the individual has been viewed as an autonomous agent in the economic system and hence crucial to understanding people's economic behaviour. It was the individual's striving to make decisions, which were best for himself/herself, which would also be best for society in the long run. This economic ideology has been elaborated on by many others. One of the most prominent economists that followed in the footsteps of Adam Smith is perhaps Paul Samuelson who has been called the 'Father of Modern Economics'. He is the author of the influential textbook, *Economics*, which is perhaps the most popular economics textbook of recent decades.[16] The book was first published in 1948 and has influenced generations of economists. The point of discussing Adam Smith and Paul Samuelson here is to call attention to the fact that they viewed the individual consumer as an autonomous decision-maker regarding his/her own purchasing choices. However, as discussed above, the use of Big Data, AI and robotics to predict consumer behaviour suggests that purchasing decisions are no longer made by autonomous consumers but by AI applications. In other words, intelligent algorithms and Big Data develop models not only of our past purchasing behaviour but also of our future purchasing behaviour. In this context, it seems reasonable to say that the idea that the consumer is an autonomous decision-maker regarding his/her own consumer choices, as proposed by Smith and Samuelsson, is no longer viable. The autonomous decision-maker has been transformed into a quantity of data that can be utilized by an intelligent robot, which makes the decisions for us.

If this assumption is correct, then the basic idea that people are autonomous decision-making agents in economic systems will also become history. One might say that the new age is transforming people into quantities of data within huge databases so that their individual actions will have no autonomous value, because they are structured into pre-designed models that determine how they will act.

Being part of a large quantity of data may sound abstract and alienating. It is definitely abstract, and it will also be almost impossible for people to understand. The individual consumer will believe that he or she acts completely autonomously and freely when making purchasing decisions. However, this perceived freedom is a constructed form of freedom, because the quantities of data and information the intelligent robots have about us imposes significant restrictions on our freedom to make our own decisions. This theoretical understanding can be expressed by saying that there is a qualitative difference between having a model of something and developing a model in order to achieve something. The first model uncovers purchasing behaviour, while the second model directs and controls purchasing behaviour.

If many of our future situations are conditional on us making the right choices, then we have reached a point in our common knowledge development where intelligent robots make these choices for us. This insight, if it is correct, could have huge consequences for the individual, for our workplaces and for the social system we are part of.

Intelligent robots, artificial intelligence, gene-editing, Big Data and other innovations will not only affect the workplace of the future, they will also drive economic growth (Baumol, 2004). In this context, William Baumol, in agreement with Joseph Schumpeter, has emphasized that innovations drive economic growth in capitalist economies.

Entrepreneurs allocate resources and create businesses. However, entrepreneurs do not develop innovations to any great extent. It is the emergence of innovations that drives economic development and economic growth. The innovators do not operate in a vacuum. They are dependent on the institutions that frame their creative power. These institutions can be framework conditions created by the government, or, as in China, around the end of the 1970s, the development of specialized zones where foreign investment could create businesses (Vogel, 2011). The point we want to make here is that innovations are created in an interaction between the creative innovators and the institutional framework conditions developed by the governing authorities. In some cases, it is the free market that is the main driver of innovations. At other times, it may be the state that single-handedly drives forward innovations (Mazzucato, 2013, 2019, 2021). As a general rule, it seems to be co-creativity between innovators, entrepreneurs and public institutions that enables the establishment of an institutional framework that fosters innovation and economic growth. It is also possible that it is the continuous change of the interactions between individual freedom to create public institutions and the political superstructure that develops innovations and entrepreneurship in the most appropriate way. The idea behind this continuous change strategy is the following proposition: If one is to achieve stable progress towards a goal, changes to the sub-elements are a prerequisite for achieving this goal. Analogously, the tightrope walker has to constantly change the position of his/her arms and legs to remain 'stable' on the tightrope so as to reach the other side.

The discussion above questions Baumol's idea that it is the free market that is the most important for creating innovations. This may be true in many instances.

However, we believe that it is the continuous change and interactions of the three conditions mentioned above that create economic growth.[17] Sometimes the state can be the most important player in the creation of innovations that promote economic growth. There are many examples of where this is the case.[18] However, it is both the private and the public sector that generate the innovations that create economic growth. When private individuals act as entrepreneurs, they are motivated, says Baumol (2004: viii), by: 'prospects of wealth, power and prestige. This is probably correct in the case of private sector initiatives. When the state functions as an entrepreneur and/or innovator, there are completely different factors that motivate the same activities. A synthesis of what drives the political system to engage in innovation and entrepreneurship can be summarized as prosperity and security for the population (Johannessen, 2022, 2023a).

Regardless of whether it is the free market or the state through an institutional framework that promotes economic development, it is the continuous innovation processes that create the basis for economic growth and prosperity for the population. How the relationships between the three factors mentioned above take shape will, in such a context, be decided pragmatically.

Baumol (2004: ix) interestingly states: 'firms cannot afford to leave innovation to chance … management is forced by market pressure to support innovative activity systematically and substantially'. What Baumol is saying here is that it is precisely innovation that is decisive for the survival of firms. At the same time, he says that this innovation process should be developed by management in a systematic way. In this context, we completely agree with Baumol. It is the prerequisites for the innovative development that we disagree with Baumol on. We believe that Baumol places too much emphasis on the innovation processes in the free market and gives too little emphasis to the public institutional framework in which many of these innovations are created.

Theoretical reflections related to innovation

Much has been written about the intelligent functions and processes of artificial intelligence and intelligent robots.[19] However, the prerequisite for these technologies to function intelligently is that there are creative processes that can develop the technology. In other words, the creative processes are intelligent in themselves. If this logic holds true, then we should focus to a greater extent on how creative processes are developed.

We are largely taught to believe that social structures have been developed so that they can function rationally. However, it is conceivable that rationality is not the sole driver behind the development of social structures. It has been shown through many historical processes that power and not rationality is the important factor that determines the development of social structures. Historical examples of power having led to the development of social structures can be found in Machiavelli (2003); Kissinger (1995, 2012, 2015; Kissinger et al., 2021). For example, it is not rationality which governs the development of feudal structures in social systems (Kotkin, 2020). The argument behind the development of feudal structures is that those people who have power are reluctant to relinquish this power. Therefore, social processes are developed that benefit the few, but which do not necessarily promote the well-being of the many.

There seems to be growing interest about how creative processes are related to intelligence in working life (Johannessen, 2022c). This interest may be limited by the increasing development of feudal structures (Kotkin, 2020; Johannessen, 2023). Creative processes can become unstable when social structures become more and more closed with a focus on protecting the power of the few. If the feudal structures fail to promote creative processes, then there is much to suggest that innovation will be hindered.

Today, we are witnessing a growing relationship between an increased degree of feudal structures and a diminution of innovation processes (Johannessen, 2023, 2024a). It seems reasonable to claim that this development increases the likelihood that the feudal structures will become more closed and isolate themselves from the outside world. In the long run, such a development will mean that the social systems that use feudal structures to protect the power of the few will not be able to compete in the open market. Having said this, it is conceivable that such feudal structures can survive for a long time.

One can be both optimistic and pessimistic about the development of future workplaces. On the one hand, social systems may develop that are based on open creative processes in social structures that promote rationality and free competition. On the other hand, feudal structures may develop which only protect the power of the few. These feudal structures will find it difficult to create enthusiasm and motivate open creative processes. Of course, we do not know whether it is the creative processes or the feudal structures that will become more prominent in the future. What we do know, however, is that if the feudal structures become prominent in many social systems, both authoritarian and democratic, then it seems reasonable to assume that the development of creative processes leading to new innovations will be hindered.

The intelligent and rational development of social systems is dependent on the ability to foster creative processes that can generate inventions. In addition to these processes, it is crucial that these inventions become innovations in a market. In order for an invention to become an innovation, it also requires people with the necessary competence who are able to distribute the innovation in a market, so it becomes of practical use. Against this background, the closed feudal structures will not only prevent the development of creative processes, but the development of competences will also be inhibited.

The new technological innovations will have the following characteristics (Ellul, 1967, 2018).

1. They are rational. This makes the processes logical and instrumental. The processes can therefore be analysed by examining the input process (transformation) and output.
2. They all have an element of artificial intelligence driving the processes.
3. They automate work processes.

4. They increase labour productivity.
5. They are designed to promote economic growth.
6. They can be operated independently, but as a general rule they are designed to be part of a larger technological system, e.g., in the global economy.
7. They permeate most economic systems in the global economy. This applies to production, distribution and consumption.
8. They are used in communication systems.
9. They are important for the development of competence.

Ellul[20] (1967, 2018) emphasizes that this technology operates on an inorganic as well as an organic level. The new technological innovations will impact the economic, social, political and cultural systems. In this way, the technological system can be understood to be on a meta level in relation to the other four systems. It is in such a context that one can say that the technological system is not only distinct from the four systems but also linked to them.

How social systems are structured influences the development of technological innovations, but technological systems also influence the development of social structures. In other words, there is an interactive relationship between technology and social systems. Against this background, we can assume that technological innovations may contribute to supporting the development of social structures that maintain and reinforce the position and dominance of power in society. With this understanding, the new technological innovations can facilitate the development of feudal structures. This development towards a form of feudal capitalism may be one of the most prominent features of the robot shock (Johannessen, 2023, 2024a).

Based on the above narratives, description, analysis and theoretical reflections, we have developed a typology for the workplace of the future based on the relationship between creative processes and feudal structures (Figure 4.2).

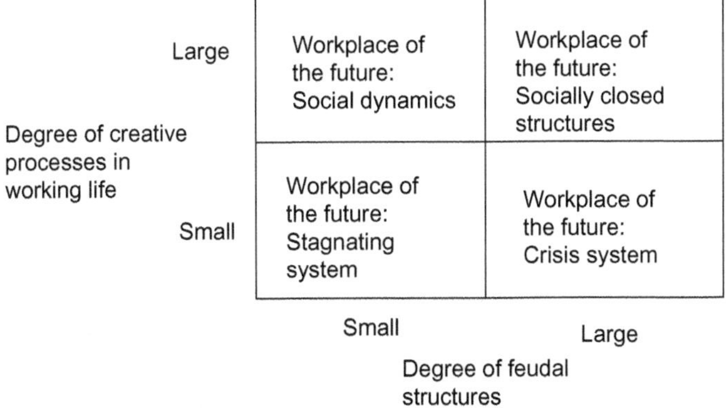

Figure 4.2 The workplace of the future showing the relationship between creative processes and feudal structures: A typology.

Practical utility related to innovation

Economic history informs us that innovations can lead to creative destruction (Aghion et al., 2021), resulting in the destruction of 'old' jobs and 'old' skills. The introduction of new technology or new institutional innovations reduces the demand for 'old' skills. After a period of time, new jobs will be created and new skills will be in demand. Economic history also informs us that technological innovations lead to increased productivity and increased automation (Frey, 2019, 2020).

Creative destruction processes change the composition of the labour force, both in terms of skills and numbers. If we consider the development from the First Industrial Revolution from the 1750s up until today's situation, as we move towards the Fourth Industrial Revolution (Schwab, 2016, 2018), many more new jobs have been created than those jobs that have been destroyed (Ford, 2016, 2021). Against this background, we can say that innovation is like Janus with two contrasting faces regarding jobs and employment. First, one of the faces shows old jobs being destroyed, because some methods and skills have become outdated and redundant. The second face shows that many new jobs and skills are created as a result of the introduction of new innovations. The crisis that these innovation processes lead to is related to the time it takes from the start of a destruction process to when the creation process manages to create new jobs and new skills.

The creative processes result in crises, and, at a later stage, new opportunities for many. It is this 'Janus economy' that is an important driving force for economic development and growth. The market decides which skills are in demand and which are not. This results in a crisis for some, while it creates career opportunities for others. However, not everyone has the opportunity to learn new skills. This will depend on many factors, such as cognitive limitations, age, educational achievements and so on. The time lag between the destruction of old skills and jobs and the creation of new skills and jobs represents a window where people and governments can adapt to the new situation. Different governments will adopt different policies. Some governments will adopt a laissez-faire attitude, leaving things to take their own course, without 'interfering'. In other words, leaving those who are negatively affected by creative destruction processes to manage for themselves. Other governments will adopt policies aimed at supporting those negatively affected by providing re-training, financial resources and so on, so they will be able to manage a less painful transition from the old to the new. At the individual level, those people who quickly manage to adapt to the new technology will cope the best.

The development of economic crises, as a result of innovations, is due to creative processes. In these creative processes, both competence development and the entrepreneurial system operate together in order to adopt the new methods that arise from the innovation process. This is an evolutionary process. The first three industrial revolutions and the imminent Fourth Industrial revolution can more suitably be viewed as 'evolutions' rather than revolutions. The reason is the long time it takes from when a new major innovation, such as the steam engine, enters the market, until it has fully penetrated the entire market, creating new jobs and skills.

Karl Weick (1979: 252) suggested that the creative destruction process could be understood as a selection, retention-enactment process. When an innovation enters the market, a selection process takes place. In other words, some elements of the economy are preserved (retention), while new elements are created (enacted). After a certain period of time, that which is created will result in new career opportunities and new entrepreneurial systems.

These evolutionary systems at the micro level are related to creative processes in an organization. In the individual workplaces, this can result in connecting several established processes, thereby creating something new. In this way, a perception will develop in the organization that the evolutionary process is democratic and developed by the employees. The contrast to this is when employees have to accept the new innovations that enter the market without active engagement. From a purely psychological point of view, it seems reasonable to assume that those who feel they are involved in developing innovations will be more enthusiastic and motivated in adopting the new technology than those who simply have to accept a decision made by the management.

The direction of the technological innovations and the patterns that emerge in social systems as a result of the new technology can be related to increased productivity and the automation of work processes.[21] What appears to be a new development regarding artificial intelligence is that this technology can come to dominate biological processes to a far greater extent. This is described by Ellul as the 'technification of biology and a simultaneous commercialization of this 'technification' (Jeronimo et al., 2013: 5). If we connect the possibilities of artificial intelligence regarding biological processes, such as gene-editing, then a picture of a future bio-technological development emerges that will affect most social systems. The further connection of these technologies to nano-technology and information technology gives us some clear indications of a developmental trend towards a society characterized by technology. This society is not unlike the society Ellul (1967, 2018) described in his books. According to Ellul, this is a society characterized by strong links between science, technology and commercial interests. This development has increasingly manifested itself half a century after Ellul first described these developmental trends. In other words, technology, science and commercialization are increasingly permeating the social space. This society is characterized by technological innovations, productivity, automation, commercialization and where profit is put before people's welfare. In such a society, the social barriers between the various strata of society will be further cemented. This can be explained by the fact that the stratum of society that mainly governs innovation processes and productivity, that is, those who lead this commercialized society, will demand that their interests are protected, and not threatened by the many who have insecure and poorly paid jobs. This will stimulate the development of feudal structures, because those who profit most from this commercialized and technological society will be interested in protecting the interests of those who generate profits. We are already witnessing the rise of societies characterized by elites and a link between economic power and meritocratic power.[22] In such a development, the social bulkheads become locked so that social mobility is no longer possible, as it was previously in capitalist systems. Feudal capitalism (Johannessen, 2023) and feudal capitalist ideology (Johannessen, 2024a) create new opportunities for a minority but few career opportunities for the majority. It is in this context one can refer to the title of Ainley's article, 'Follow your dreams and attend to universities if possible', as being non-valid, and the title of his book, *Betraying a Generation: How Education is Failing Young People* (2016), as hitting the nail on the head; that is, higher education no longer guarantees that graduates will be able to find secure and well-paid jobs. However, we can modify Ainley's view by suggesting that a small elite from the technological universities and other universities that focus on commercializing innovations will reach the top of the feudal structures. Ainley's point, which we support, is that many university graduates will find it difficult to fulfil their dreams, expectations and hopes regarding future well-paid and secure jobs. Many of these graduates will end up in what Standing (2014, 2014a, 2016, 2019) terms the precariat. However, the precariat is not a homogenous social group. It is divided into several layers (Johannessen, 2020, 2020a, 2020b). These layers can be viewed as feudal structures that contribute to the development of economic and social barriers, both vertically and horizontally. At an overall level, this will result in a reduction of social dynamics. It is against this background that we can say that the technological innovations we see emerging will create new feudal structures and at the same time hinder the social dynamics of social systems.

Based on the narratives, description, analysis, theoretical reflections and the review of the utility value, we have developed a Boudon-Coleman diagram for the relationship between innovation and the workplace of the future Figure 4.3).

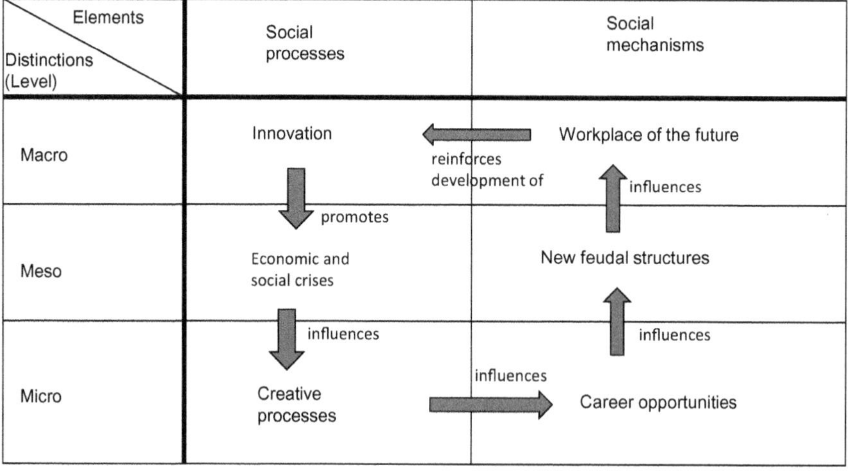

Figure 4.3 Aspects of the relationship between innovation and the workplace of the future: A Boudon-Coleman diagram.

Systemic connections related to innovation

There are systemic connections between four elements that we will elaborate on here and how they affect each other. These elements are: innovation, automation, productivity and employment. The systemic connections between these elements will affect the workplace of the future in different ways.

Innovation

Innovation has for a long time been the driving force behind economies.[23] Innovation does not have to be something that is developed in an individual enterprise. New technology can be purchased via licenses and in this way promote the economic competitiveness of businesses. However, the situation is different if it is a nation that prevents the sale of new technology to other nations. For instance, this occurred as far back to the First Industrial Revolution, when Britain placed restrictions on the sale of new technology to other countries (Allen, 2009; Mantoux, 1961). We can see a similar development today. The Chips and Science Act (9 August 2022) provides subsidies for chip manufacturing in the US, and funding for semiconductor research. Moreover, the US has prohibited the sale of high-tech equipment to China that can be used in the development of artificial intelligence.[24] In other words, the Act aims to boost the competitiveness of the US semiconductor industry in relation to China. The Act will have a direct impact on the development of high-tech jobs in the US , by creating more jobs for the STEM workforce. The Act also aims to create regional high-tech hubs, a bigger STEM workforce and increase STEM workforce training and development.[25] Some commentators have raised concerns regarding the protectionist provisions of the Chips and Science Act with regard to US government subsidies to high-tech industry. One might go as far to say that the Act is the final manifestation that free global competition regarding high technology has come to an end. In other words, after the passing of this Act, a different type of competition now exists between the two largest economies in the world, China and the US.

There is little doubt that this Act was a direct result of the growing technology competition between the US and China. We have previously seen this development regarding restrictions to free competition in technological industries in the Huawei case. The US banned Huawei from developing 5G networks in the US, and they also managed to convince some of their allies that Huawei posed a security risk, so Huawei was also banned in some European countries.

In this way, we can say that innovation, at least from the 1750s onwards towards the Fourth Industrial Revolution, has been used as a weapon' to prevent the economic development and growth of competing countries. The Chips and Science Act also aims to strengthen the US military. Amongst other things, if China led the development of AI, this could pose a danger to the US military superiority.[26]

The debate regarding the Chips and Science Act has mainly focused on the US economy and jobs. It is believed that if the US can prevent the sale of chips and other high-tech products to China, then this will help American jobs. On the other hand, this Act will probably also restrict the creation of Chinese high-tech jobs.

In other words, the idea is that American jobs will benefit from restrictions to free competition in high-tech industries.

Using national security arguments in the development of high technology, the US is able to give an advantage to its own entrepreneurs, because they are able to develop businesses without foreign competition.

The innovative entrepreneurs, in contrast to the traditional entrepreneurs, promote economic development based on high-tech innovations. On the other hand, the traditional entrepreneurs fill gaps in the market and promote balance in the economy. Innovations, and the innovative entrepreneurs who promote innovations, however, create gaps in the market and promote imbalance in the economy. It is this dynamic relationship between the traditional entrepreneurs and the innovative entrepreneurs that creates economic development and growth.

In order to promote innovation that drives the labour market, there is an absolute prerequisite that must be present. This prerequisite is the development and growth of a large STEM workforce.

Automation

Automation can be understood as routinizing technological innovations so they become incorporated into the normal organizational activities. When this routinization occurs, the technological innovations, minor or major, are first used in the existing organization of the production process. This often means that productivity is not as great as it could be using the new technology. Gradually, however, the new technology is adapted to new forms of organization, such as institutional innovations. The new technology, in combination with institutional innovations, results in large increases in productivity. This process takes time, from the time the technology is introduced into the existing organization of the work processes, until the work processes are changed. Consequently, because of the time lag, the new technology is not experienced as a crisis by those who are involved in the automation process. Both workers and businesses have time to adapt so that the new technology is not perceived as a threat.

There is a time lag between the introduction of new technology and an optimal application of the technology; in other words, it takes time for the automation process to be fully implemented. This sluggishness in the automation process is nothing new. It has happened in most cases where new technology has become part of an established work process (Ellul, 1967, 2018). It is precisely this sluggishness in the automation process that means that innovation processes both maintain and change social systems at the same time. First, the innovations maintain the established work organization and thus give workers and businesses time to adapt. Next, automation changes many of the work processes in relation to how they were previously organized. It is this process, where innovations both act as an anchor and a driver of economic growth that characterizes the automation process. Figuratively, this can be compared to revving up a car engine with the handbrake on and then releasing the brake resulting in dramatic acceleration.

If we consider new businesses that start up on the basis of new technology, we see completely different automation processes than in the case of established businesses that use the same technology. The new businesses automate from day one the work processes that take a long time to optimize in the established businesses. The time lag factor can often create problems for established businesses. This occurs when new businesses emerge that are in direct competition with the established businesses. Perhaps the most illustrative case in this context is the competition that took place between the established Kodak company and the new business, Instagram.[27] One can illustrate the behaviour of the established businesses, or the lack of action, by referring to the analogy of the 'basking frog' (Handy, 2002, 2007, 2016). Established companies have long been used to the fact that everything will turn out okay if they just wait a while. This is what the basking frog believed too. The frog was put into a cooking pot filled with cold water. The water was heated up extremely slowly. The frog acclimatized to the gradually warming water and eventually basked itself to death! Nokia, like Kodak, was unable to adapt to changing conditions. The company was outcompeted with the introduction of smartphones, and they lost their leading position in the industry when they were unable to adapt to technological changes.

Automation processes are most visible between established businesses and new innovative entrepreneurial businesses entering the same market. This is when the contested market space can run red with the blood spilt through competition. The established companies either change quickly or they are pushed out of the market, as happened with Kodak and Nokia.

How effective automation is when new technology enters the market depends on the degree of compatibility between: 1. The new technology, 2. Organization, 3. Expertise and 4. other adaptations that must be made in order to optimize the use of the new technology. It is precisely the optimization of technology in relation to the above which affects the development of productivity.

Productivity

It is the entrepreneurs within an enterprise (intrapreneurs) and entrepreneurs in other enterprises who drive economic development through changing established procedures and/or adopting innovations. When an entrepreneur or intrapreneur uses innovations, they are called innovative entrepreneurs (Spulber, 2014). It is these innovative entrepreneurs who create an imbalance in the market, and thus challenge the economic and social dynamics. When the innovative entrepreneurs, for one reason or another, encounter obstacles in their application of innovations, productivity slows down. These inhibiting factors may be institutional constraints, for instance, new laws that make it more difficult to start new businesses. The reasons given for such bureaucratic obstacles are, as a general rule, related to quality and overall safety for the social system. However, for the innovative entrepreneur, such laws are often just viewed as obstacles, regardless of the reasons given. If productivity is inhibited for any reason, then this affects the social system. In its consequences, this leads to a reduction in economic growth.

Productivity growth is directly related to economic growth (Black, 2013). The question then becomes: What drives productivity growth? The answer, which has increasingly been accepted as a fact, can be expressed in the following proposition: Productivity growth is driven to a large extent by innovation (33 per cent). This proposition is derived from the research of Jorgensen and Fraumeni (1987) as well as Baumol (2004). In addition to innovation, the institutional framework within which businesses operate is important for productivity development. This insight is largely the result of the research of Douglass North, the Nobel Prize winner in Economic Sciences.[28]

The proposition that emerges from the above reasoning is: Productivity growth is driven by innovations and also dependent on the institutional framework. We know that internal and external entrepreneurs ensure the practical implementation of productivity growth by adopting innovations within existing institutional frameworks; the question is: Which institutional framework is it that inspires the entrepreneur to make that little extra effort in developing innovations. The answer is given by Baumol (2004: 60): 'It is my belief that the explanation for this rise and fall of entrepreneurial activity is grounded in simple dollars and cents'. In plain text, Baumol says that the risks the entrepreneur takes is weighed against the profit that lies in the opportunities that are available. The same insight also became evident in communist China, when Deng Xiaoping opened up China to Western capital and initiated the special economic zones (SEZs) along the Chinese east coast (Vogel, 2011). What Xiaoping stated quite explicitly was that some people had to become rich before others (Deng, 1994).

In an input-output process (Leontief, 1979, 1986), the entrepreneurs represent input; innovations and institutional frameworks represent the transforming force; while productivity represents output. In other words, this constitutes the four most important dimensions of economic growth: Innovative entrepreneurs, the institutional framework, innovation and productivity. The relationship between the innovative entrepreneurs and the institutional framework is important to understand, because this drives the economic dynamics at a micro level. An important connection here is as shown above. It is the possibility of profit in relation to risk that drives the entrepreneur. As we have shown above, this applies to both capitalist and communist economies.

It is important to be aware of the fact that profit is not the only factor that drives the entrepreneur. North has indicated an equally important institutional driving force, which can be related to Asplund's motivation theory[29] (Asplund, 2010). In brief, this theory can be described in the following way: *People are motivated by social responses* (Asplund, 2010: 221–229). Asplund points out that: *When people receive social responses, their level of activity increases.* Asplund's motivation theory is consistent with North's Action Theory (North, 1990, 1993). Understood in this way, it seems reasonable to connect the two theories in the statement: *People are motivated by the social responses rewarded by the institutional framework.* This proposition expands our understanding of what drives the entrepreneur. Profit is important, but there are also other types of rewards that drive the entrepreneur.

The labour force

The workforce's level of competence and the composition of the workforce's competence will directly influence the development of the workplaces of the future.

It can be said that innovation and the workforce mutually reinforce each other. Innovation leads, amongst other things, to greater productivity. In its historical consequences, this has led to increased income for the workforce as a whole. Another consequence is that the level of education and the state of health of the population will improve qualitatively as a result of economic growth. Economic growth also leads to an increase in life expectancy, due to improved healthcare, amongst other things (Frey, 2019, 2020). Moreover, when people's skills and expertise reach a higher level, this positively affects the development of innovations.[30]

As mentioned above, economic and social systems are driven forward by innovation. Marx views the struggle between the social classes as the driving force of history. This may be correct, but, in this context, it is perhaps advantageous to make a distinction between political history and economic history. The dynamism of economic history is largely linked to innovation. Political history can be understood as changes in power relations. It is against this background, and from this perspective, that one can say that economic history is driven forward by various types of innovation.

Without the introduction of innovations in an economy, this will, in the long run, affect the workforce negatively. Amongst other things, innovations are directly linked to automation and increased productivity. There is also a direct link between innovations and workforce development. For instance, this can be related to institutional innovations, market innovations, new ways of organizing, etc.

The spill-over effect from innovations to the workforce is what has the greatest influence on the development and change in the workforce's level of competence, as well as the direction of this competence. It is when the competence is compatible with the innovations found in the market that it can be said that there is an optimal relationship between innovations and the workforce.

The proposition that emerges from the above reasoning is: The lower the level of innovation processes in an economy, the more probable it is that most people will be worse off. The rationale is that without innovations there will be little dynamism and development in the economic system. Without various types of major innovations, society would be characterized by a struggle for survival, which was a feature of the pre-industrial age. Prior to the First Industrial Revolution, most people spent most of their money on basic foodstuffs (Baumol, 2004: 135). Right up until the beginning of the 19th century, European countries were constantly threatened by periods of famine (Palmer, 1964: 49). It was the various types of major innovations introduced during the First Industrial Revolution that led to economic growth, removing the spectre of famine that hung over Europe in the pre-industrial age (Braudel, 1979: 73–75).

The spill-over effect of innovation can be affected by, amongst other things, the workforce's competence level and competence composition. The more the competence is compatible with the central innovations, the greater the probability that the

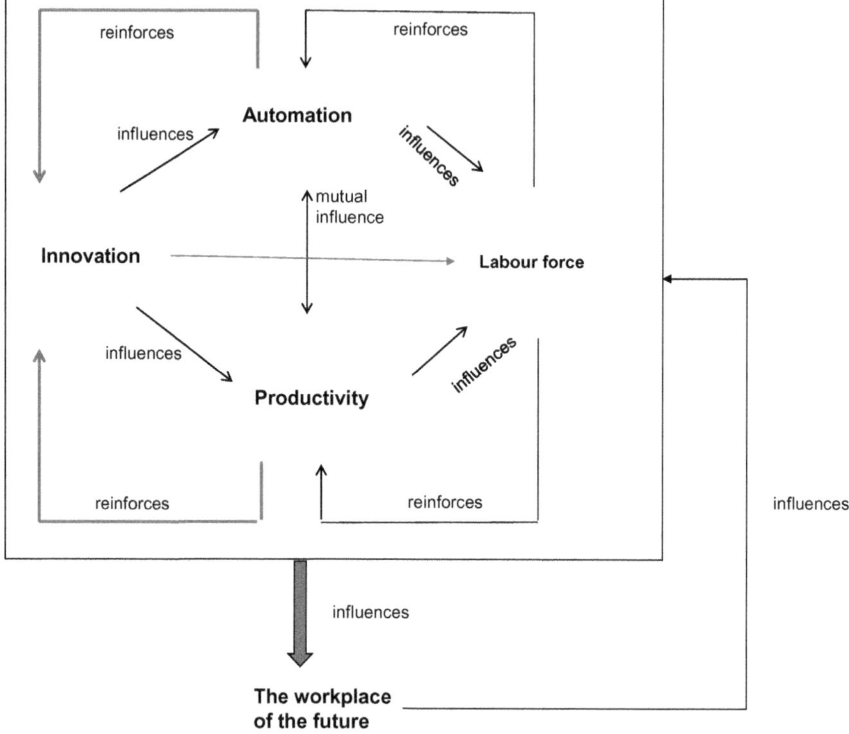

Figure 4.4 How innovation can affect future workplaces: A loop diagram.

spill-over effect will be large. A country that over time uses part of its surplus to invest in mass education and then focuses on high technological competence will, over time, gain an advantage in relation to those countries that do not. China (from 1970 until the 2020s) is a good example of this phenomenon (Jain, 2021; Drache et al., 2019).

Based on the above narratives, analysis, description, theoretical reflections, the review of the practical utility and the systemic connections, we have developed a loop diagram for how innovation can affect the workplaces of the future Figure 4.4).

Sub-conclusion related to innovation

This chapter has stressed that innovation is the most important driving force for economic growth. This idea is in complete accordance with Baumol's (2004) analysis. However, Baumol states that innovation is the most important driving force for capitalism. This is perhaps correct, but it is not only capitalist economies that are driven by innovations. China provides evidence that innovation can also drive communist economies. This is why we state here that innovation is the most important driving force for growth and development in all economic systems, regardless of which political superstructure an economic system may have. This statement is validated by historical evidence in Western countries and in communist countries, such as China (Hamel, 2008, 2012; Johannessen, 2021, 2022, 2023a).

For millennia, innovation as an element of economic activity has been integral to production. Innovations have existed at least since the invention of the wheel in Mesopotamia in approximately 4000 BCE. However, during the feudal era, which spanned 1000 years, the nobility hindered technological agricultural innovations, because they feared the innovations would result in the serfs becoming less dependent on the nobility (Marx & Engels, 2020). Innovation processes accelerated with the birth of capitalism, which started around the early 1600s in Holland with the foundation of a stock exchange and the establishing of joint-stock companies. Innovation processes were further accelerated during the First Industrial Revolution around 1750 in Britain. While feudal society held back innovations for reasons of power politics, innovations were promoted during the capitalist era for economic reasons.

Regardless of which political systems throughout history have promoted or inhibited the development of innovations, innovations have impacted social systems and workplaces. This impact has been positive for some but negative for others. Those who have quickly adapted their skills to the new technology have benefitted from this process. Those who, for various reasons, have not adapted their skills to the new technology have suffered. There is nothing to suggest that the new technology we see developing, such as artificial intelligence, intelligent algorithms, intelligent robots, Big Data, gene-editing and so on, will not affect future workplaces any differently than previous innovations have done. However, we can modify this view by saying that the new technology will very likely change the workplaces of the future more radically than previous innovations. This can be explained by the fact that artificial intelligence is a so-called General Purpose Technology (GPT), which will impact most work activities and processes as we move towards the Fourth Industrial Revolution.[31] GPTs affect the entire economic system (Lipsey & Carlaw, 2005). The historical examples of GPTs are the steam engine, electricity, the internet, rail transport, information technology and the computer. All these innovative technologies promoted productivity and led to automation processes in social systems.

In this chapter, we have focused on how innovation will affect the workplaces of the future. This could also have been formulated so that it was related to how innovation affects the whole of the social system. Our point here is that when workplaces are affected and changed, the entire economic system changes as well. The general answer we have given is that workplaces will be automated over time to the point of being completely unrecognizable. The workplaces are also related to the microeconomic aspect of the economic system, and this is what we have focused on in this chapter and this book, although we have also considered some macroeconomic and global tendencies of the technological innovations, as we move towards the Fourth Industrial Revolution.

History does not repeat itself. However, history sends signals to decision-makers in the present. These signals then manifest themselves in the choices that are made. It is these choices that can affect the future. This is also the case with innovations. The technological innovations we see developing today, such as intelligent robots, will be manifested in relation to decisions concerning productivity, automation and new ways of organizing production, distribution and consumption processes. In this way, the workplaces of the future are influenced not only by innovations that are being developed today but also by the signals that history sends to today's decision-makers about, amongst other things, previous innovations and their consequences for social systems.

Notes

1 Acemoglu & Robinson, 2006; Acemoglu et al., 2015; Frank, 2004, 2020; Carney, 2020; Lind, 2020; Reich, 2009; Hertz, 2002; Iversen & Soskice, 2019; Levitsky & Ziblatt, 2019; Przeworski, 2019; Rachman, 2022; Rosanvallon, 2006; Ziblatt, 2017.
2 McKinsey report: https://www.mckinsey.com/~/media/mckinsey/featured%20insights/Digital%20Disruption/Harnessing%20automation%20for%20a%20future%20that%20works/MGI-A-future-that-works-Executive-summary.ashx
3 https://en.wikipedia.org/wiki/Jacques_Ellul
4 Antonelli, 2009; Atkitson & Ezell, 2014; Brynjolfson & Saunders, 2013; Billington, 2020; Chesbrough, 2003; Hamel, 2002, 2007, 2008, 2012; Hamel & Zanini, 2020; Johannessen, 2018, 2020, 2020a, 2020b, 2020c, 2020d, 2021, 2021a, 2021b, 2022, 2022a, 2022b, 2022c, 2022d; OECD, 2001; Skarzynski & Gibson, 2008; Tucker, 2002; Unterberg, 2013; Von Hippel, 2005.
5 Li, 2020; Johannessen, 2020a; Schwab, 2016, 2018; Skidelsky & Craig, 2020.
6 Acemoglu et al., 2015; Atkinson, 2015; Dorling, 2015, 2019; Eubanks, 2017; Milanovic, 2016; Picket, 2017; Piketty, 2014, 2016; Stilwell, 2019; Weeks, 2014.
7 https://en.wikipedia.org/wiki/General-purpose_technology
8 https://www.brookings.edu/articles/the-solow-productivity-paradox-what-do-computers-do-to-productivity/
9 https://www.mckinsey.com/capabilities/mckinsey-digital/our-insights/is-the-solow-paradox-back
10 https://en.wikipedia.org/wiki/Productivity_paradox
11 Brown, 2018; Beeson & Li, 2016; Fannin, 2019; Ferdinand, 2016; Berli & Hung, 2020; Jain, 2021; Johannessen, 2021, 2023a; Li, 2021; Rocca, 2018; Rozelle, 2020; Vangeli, 2019.
12 First published in 1932.
13 First published in 1949.
14 Berlinski, 2000; Bonaccorso, 2020; Esposito, 2022; Mirjalili, 2019; Szeliski, 2021; Zelinka & Guanrong, 2019.
15 https://en.wikipedia.org/wiki/Adam_Smith
16 https://da.wikipedia.org/wiki/Paul_Samuelson
17 1. The individual freedom to create, 2. Public institutions and 3. The political superstructure. It is the continuous change of the interaction and composition of these three elements that creates innovations and economic growth, which is our thesis.
18 Johannessen, 1990, 2018, 2020, 2020a, 2020b, 2021, 2022, 2023, 2023a.
19 Bohr & Memarzadeh, 2020; Carou et al., 2022; Crevier, 1993; Esposito, 2022; Johannessen, 2020b, 2021, 2021a, 2023; Kalia, 2016; Kumar et al., 2021; Lidstrømer & Ashrafian, 2022; Li, 2020; Mahajan, 2021; Russel & Norvik, 2016; Simon, 2019; Xing et al., 2020.

20 https://en.wikipedia.org/wiki/Jacques_Ellul
21 Bessen, 2020; Ford, 2016, 2021; Frey, 2019, 2020; Goldin & Katz, 2008; Kotkin, 2020; Kumar et al., 2021; Sachs, 2020; Susskind, 2018; Susskind, 2020; Szeliski, 2021.
22 https://en.wikipedia.org/wiki/Meritocracy
23 Baumol, 2004; Nelson & Winter, 1982; Schumpeter, 1951, 1954, 1989, 2002, 2006, 2010; Teece, 1988.
24 https://www.mckinsey.com/industries/public-and-social-sector/our-insights/the-chips-and-science-act-heres-whats-in-it
25 STEM-science, technology, engineering, mathematics.
26 https://en.wikipedia.org/wiki/CHIPS_and_Science_Act
27 https://www.mcorpcx.com/articles/instagram-vs-kodak-smart-customers-sidestep-stupid-companies
28 North, 1981, 1990, 1993, 1996, 1997. (https://en.wikipedia.org/wiki/Douglass_North)
29 Asplund's motivation theory, a term we use here, is based on Asplund's research..
30 Antonelli, 2009; Bessen, 2020; Brynjolfsson & McAfee, 2011, 2014; Ford, 2016, 2021; Goldin & Katz, 2008; Habakkuk, 1962; Lesavich, 2001.
31 https://en.wikipedia.org/wiki/General-purpose_technology

References

Acemoglu, D. & Autor, D.H. (2011). Skills, tasks and technologies: Implications for employment and earnings, in Card, D. & Ashenfelter, O. (eds.), *Handbook of labor economics*, vol. 4, Elsevier, Amsterdam, pp. 1043–1171.

Acemoglu, D. & Robinson, J.A. (2006). *Economic origins of dictatorship and democracy*, Cambridge University Press, Cambridge.

Acemoglu, D.; Suresh, N.; Pascual, R. & Robinson, J.A. (2015). Democratic redistribution, and inequality (Chapter 21), In Atkinson, A.B. & Bourguignon, F. (eds.), *Handbook of income distribution*, vol. 2, Elsevier, London, pp. 1885–1966

Ackoff, R.L. (1981). *Creating the corporate future: Planned or be planned for*, John Wiley % Sons, New York.

Aghion, P.; Antonin, C. & Bunel, S. (2021). *The power of creative destruction*, Harvard University Press, Cambridge, MA.

Ainley, P. (2014). Follow your dreams and attend to universities if possible, *Latitude*, 21 December.

Ainley, P. (2016). *Betraying a generation: How education is failing young people*, Policy Press, Bristol.

Allen, R.C. (2009). *The British industrial revolution in global perspective*, Cambridge University Press, Cambridge.

Amabile, T. (1996). *Creativity in context*, Westwiew Press, New York.

Antonelli, C. (2009). The economics of innovation: From the classical legacies to the economics of complexity, *Economics of Innovation and New Technology*, 18 (7): 611–646. DOI: 10.1080/10438590802564543. To link to this article: http://dx.doi.org/10.1080/10438590802564543.

Asplund, J. (2010). *Det sociala livets elementära former*, Korpen, Stockholm.

Atkinson, A.B. (2015). *Inequality: What can be done*, Harvard University Press, Cambridge.

Atkinson, R.D. & Ezell, S.J. (2014). *Innovation economics: The race for global advantage*, Yale University Press, New York.

Baumol, W.J. (2004). *The free market innovation machine: Analyzing the growth miracle of capitalism*, Princeton University Press, Princeton, NJ.

Beeson, M. & Li, F. (2016). China's place in regional and global governance: A new world comes into view, *Global Policy*, 7 (4): 491–499.

Berlie, J.A. & Hung, S. (2020). The greater Bay area and the role of Hong Kong and Macau SARs in the Belt and Road initiative, in Berlie, J.A. (ed.), *China's globalization and the belt and road initiative*, Palgrave, London, pp. 77–100.
Bessen, J. (2020). Attitudes to technology: Part 1, in Skidelsky, R. & Craig, N. (eds.), *Work in the future*, Palgrave, London, pp. 83–88.
Billington, D.P. (2020). *From insight to innovation*, MIT Press, Boston.
Berlinski, D. (2000). *The advent of the algorithm*, Harcourt Books, New York.
Black, S.W. (2013). *Productivity growth and the competitiveness of American economy*, Springer, London.
Bohr, A. & Memarzadeh, K. (2020). *Artificial intelligence in healthcare*, Academic Press, New York.
Bonaccorso, G. (2020). *Mastering machine learning algorithms*, Packt Publishing, London.
Braudel, F. (1979). *Civilization and capitalism, 15*th *to 18*th *century*, vol. 1, Harper & Row, New York.
Brown, K. (2018). *Chinas dream*, Polity, London.
Brynjolfsson, E. & McAfee, A. (2011). *Race against the machine*, Digital Frontier Press, New York.
Brynjolfsson, E. & McAfee, A. (2014). *The second machine age*, W.W. Norton & Company, New York.
Brynjolfsson, E. & Saunders, A. (2013). *Wired for innovation: How information technology is reshaping the economy*, The MIT Press, London.
Carney, T.P. (2020). *Alienated America*, Harper, New York.
Carou, D.; Sartal, A. & Darim, J.P. (2022). *Machine learning and artificial intelligence with industrial applications: From big data to small data*, Springer, London.
Chesbrough, H.W. (2003). *Open innovation: The new imperative for creating and profiting from technology*, Harvard Business School Press, Boston.
Crevier, D. (1993). *AI: The tumultuous search for artificial intelligence*, BasicBooks, New York, NY.
Davies, K. (2021). *Editing humanity*, Pegasus Books, London.
Deng, X. (1994). *Selected works of Deng Xiaoping, vol III (1982–1992)*, Foreign Language Press, Beijing.
Dorling, D. (2015). *Injustice: Why social inequality still persists*, Policy Press, London.
Dorling, D. (2019). *Inequality and the 1%*, Verso, London.
Doudna, J. & Sternberg, S. (2018). *A crack in creation: Gene editing and the unthinkable power to control evolution*, Houghton Mifflin, New York.
Drache, D.; Kingsmith, A.T. & Qi, D. (2019). *One road many dreams: China's bold plan to remake the global economy*, Bloomsbury, New York.
Ellul, J. (1967). *The technological society*, Vintage, London.
Ellul, J. (2018). *The technological system*, Wipf and Stock, London.
Esposito, E. (2022). *Artificial communication: How algorithms produce social intelligence*, The Mit Press, Boston.
Eubanks, V. (2017). *Automating inequality*, St. Martin's Press, New York.
Fannin, R.A. (2019). *Tech titans of china*, Nicholas Brealey Publishing, London.
Ferdinand, P. (2016). The China dream and one belt one road: Chinese foreign policy under Xi Jinping, *International Affairs*, 92 (4): 941–957.
Ford, M. (2016). *The rise of the robotics, Technology and the threat of mass unemployment*, Oneworld, New York.
Ford, M. (2021). *Rule of the robots*, Basic Book, London.
Frank, T. (2004). *What's the matter with Kansas? How conservatives won the Heart of America*, Metropolitan Press, New York.
Frank, T. (2020). *People without power*, Scribe, London.
Frey, C.B. (2019). *The technology trap: Capital, labor, and power in the age of automation*, Princeton University Press, Princeton.

Frey, C.B. (2020). Attitudes towards technology: Part II, in Skidelsky, R. & Craig, N. (eds.), *Work in the future: The automation revolution*, Palgrave, London, pp. 89–97.
Goldin, C. & Katz, L.F. (2008). *The race between education and technology*, The Belknap Press, New York.
Habakkuk, J. (1962). *American and British technology in the nineteenth century, the search for labor savings inventions*, Cambridge University Press, Cambridge.
Hamel, G. (2002). *Leading the revolution: How to thrive in turbulent times by making innovation a way of life*, Harvard Business School Press, Boston.
Hamel, G. (2007). *The future of management*, Harvard Business School Press, Boston.
Hamel, G. (2008). Introduction, in Skarzynski, P. & Gibson, R. (eds.), *Innovation to the core*, Harvard Business Press, Boston, pp. xvii–xix.
Hamel, G. (2012). *What matters now: How to win in a world of relentless change ferocious competition, and unstoppable innovation*, John Wiley & Sons, New York.
Hamel, G. & Zanini, M. (2020). *Humanocracy: Creating organizations as amazing as the people inside them*, Harvard Business Review Press, Boston, MA.
Handy, C. (2002). *The age of unreason*, Arrow, New York.
Handy, C. (2007). *Myself and other important matters*, Arrow, New York.
Handy, C. (2016). *The second curve, Thoughts on reinventing society*, Random House, London.
Harari, Y.N. (2022). Introduction, in Huxley, A. (ed.), *Brave new world*, Vintage Classic, London, pp. vii–xi.
Hertz, N. (2002). *The silent takeover: Global capitalism and the death of democracy*, Arrow, London.
Hertz, N. (2020). *The lonely century: Coming together in a world that's pulling apart*, Sceptre, London.
Iversen, T. & Soskice, D. (2019). *Democracy and prosperity*, Princeton University Press, Princeton.
Jain, R. (2021). *China's soft power and higher education in South Asia*, Routledge, London.
Jeronimo,H.M.; Garcia, J.L. & Mitcham, C. (2013). Introduction: Ellul returns, in Jeronimo, H.M.; Garcia, J.L. & Mitcham, C. (eds.), *Jacques Ellul and the technological society in the 21*st *century*, Springer, London, pp. 1–17.
Johannessen, J-A. (1990). *Information management*, Ph.D. Thesis, Stockholm University, Stockholm.
Johannessen, J-A. (2018). *Innovation leads to economic crises*, Palgrave, London.
Johannessen, J-A. (2020). *The workplace of the future*, Routledge, London.
Johannessen, J-A. (2020a). *Automation, innovation and economic crises: Survival the fourth industrial revolution*, Routledge, London.
Johannessen, J-A. (2020b). *Artificial intelligence, automation and the future of competence at work*, Routledge, London.
Johannessen, J-A. (2020c). *Knowledge management for leadership and communication: AI, innovation and the digital economy*, Emerald, London.
Johannessen, J-A. (2020d). *Knowledge management philosophy: Communication as a strategic asset in knowledge management*, Emerald, London.
Johannessen, J-A. (2021). *China's innovation economy: Artificial Intelligence and the new silk road*, Routledge, London.
Johannessen, J-A. (2021a). *Artificial intelligence, automation and ethics in the innovation economy*, Routledge, London.
Johannessen, J-A. (2021b). *Communication as social theory: The social side of knowledge management*, Emerald, London.
Johannessen, J-A. (2022). *The new silk road and the innovation economy in China*, Routledge, London.
Johannessen, J-A. (2022a). *Creativity, innovation and the fourth industrial revolution: The da Vinci strategy*, Routledge, London.

Johannessen, J-A. (2022b). *A systemic approach to continuous change in the innovation economy*, Routledge, London.
Johannessen,J-A. (2022c). *Intelligent robots consciousness and creativity: The search for hidden knowledge, The cognitive side of knowledge management*, Emerald, London.
Johannessen, J-A. (2022d). *The philosophy of tacit knowledge*, Emerald, London.
Johannessen, J-A. (2023). *Feudal capitalism in the innovation economy*, Routledge, London.
Johannessen, J-A. (2023a). *De-Globalization in the innovation economy: **Technological innovations trigger conflict between China and the United States***, Routledge, London.
Johannessen, J-A. (2024a). *The ideology of feudal capitalism in the innovation economy*, Routledge, London.
Johannessen, J-A.; Olsen, B. and Lumpkin, G.T. (2001). Innovation as newness: What is new, how new, and new to whom? *European Journal of Innovation Management*, 4 (1): 20–31.
Johannessen, J-A. & Sætersdal, H. (2020). *Automation, innovation and work*, Routledge, London.
Jorgenson, F. & Fraumeni, B. (1987). *Productivity and U.S. economy's growth*, Harvard University Press, Cambridge, MA.
Kalia, P. (2016). Artificial intelligence in E-commerce: A Business Process analysis, in Bhargava, C. & Kumar Sharma, P. (eds.), *Artificial intelligence. Fundamentals and applications*, Routledge, London, pp. 9–19.
Kissinger, H. (1995). *Diplomacy*, Simon & Schuster, New York.
Kissinger, H. (2012). *On China*, Penguin, New York.
Kissinger, H. (2015). *World order*, Penguin, New York.
Kissinger, H.A.; Schmidt, E. & Huttenlocher, D. (2021). *The age of AI: And our human future*, Little Brown and Company, New York.
Kotkin, J. (2020). *The coming of neo-feudalism: A warning to the global middle class*, Encounter Books, New York.
Kumar S., Aggarwal, V. & Gupta, S. (2021). Artificial intelligence and nanotechnology: A super convergence, In Bhargava, C & Kumar Sharma, P. (eds.), *Artificial intelligence: Fundamentals and applications*, Routledge, London, pp. 1–9.
Kurzweil, R. (2005). *The Singularity is near*, Penguin, London.
Kurzweil, R. (2008). *The age of spiritual machines: When computers exceed human intelligence*, Penguin, London.
Kurzweil, R. (2013). *How to create a mind: The secret of human thought revealed*, Penguin Books, New York.
Leontief, W. (1979). Is technological Unemployment inevitable? *Challenge*, 22 (4): 48–50.
Leontief, W. (1986). *The future impact of automation on workers*, Oxford University Press, Oxford.
Lesavich, S. (2001). Are all business methods patents "one click" away from vulnerability? *Intellectual Property and Technology Law Journal*, 13 (6): 1–5.
Levitsky, S. & Ziblatt, D. (2019). *How democracies die*, Penguin, New York.
Lidstrømer, N. & Ashrafian, A. (2022). *Artificial intelligence in medicine*, Springer, London.
Lind, M. (2020). *The new class war: Saving democracy from the managerial elite*, Portfolio, New York.
Li, C. (2021). *Middle class Shanghai,: Reshaping US-China engagement*, Brooking Institution Press, Washington, DC.
Li, R. (2020). *Artificial intelligence revolution: How AI will change our society, economy and culture*, Skyhorse Publishing, New York.
Lipsey, R.G. & Carlaw, K.I. (2005). *Economic transformations: General purpose technology and long term economic growth*, Oxford University Press, Oxford.
Machiavelli, N. (2003). *The prince*, Penguin Classic, London.
Mahajan, P.S. (2021). *Artificial intelligence in healthcare*, MedMantra, London.

Mantoux, P. (1961). *The industrial revolution in the eighteenth century: An outline of the beginning of the modern factory system in England*, Routledge, London.
Marx, K. & Engels, F. (2020) (1867). *The communist manifesto*, Wordsworth, New York.
Mazzucato, M. (2013). *The entrepreneurial state*, Penguin, London.
Mazzucato, M. (2019). *The value of everything*, Pelican, London.
Mazzucato, M. (2021). *Mission economy*, Allen Lane, New York.
Milanovic, B. (2016). *Global inequality*, The Belknap Press, New York.
Mirjalili, S. (2019). *Evolutionary algorithms and neural networks*, Springer, London.
Nelson, R.R. & Winter, S.G. (1982). *An evolutionary theory of economic change*, Cambridge, MA, Belhaven Press.
Nietzsche, F. (1968). *Twilight of idols*, Penguin, London
North, D.C. (1981). *Structure and change in economic history*, Norton, New York.
North, D.C. (1990). *Institutions, institutional change and economic performance*, Cambridge University Press, Cambridge.
North, D. (1993). Nobelforedraget. http://www.nobelprize.org/nobel_prizes/economics/laureates/1993/north-lecture.html#not2, date of reading: 4.5.2021.
North, D.C. (1996). Epilogue: Economic performance through time, in Alston, L.J.; Eggertson, T. & North, D.C. (eds.), *Empirical studies in institutional change*, Cambridge, Cambridge University Press, pp. 342–355.
North, D.C. (1997). Prologue, in Drobak, J.N. & Nye, J.V.C. (eds.), *The frontiers of the new institutional economics*, Academic Press, New York, pp. 3–13.
OECD (2001). *Innovative clusters: Driving of national innovation-systems*, OECD, Paris.
Orwell, G. (2008). *1984*, Penguin, London.
Palmer, R. (1964). *The age of democratic revolution*, vol. 2, Princeton University Press, Princeton.
Picket, K. (2017). Foreword, in Brown, R. (ed.), *The inequality crises*, Policy Press, London, pp. vii–viii.
Piketty, T. (2014). *Capital in the twenty-first century*, The Belknap Press of Harvard University Press, Boston.
Piketty, T. (2016). *Chronicles: On our troubled times*, Viking, London.
Przeworski, A. (2019). *Crises of democracy*, Cambridge University Press, Cambridge.
Rachman, G. (2022). *The age of the strongman: How the cult of the leader threatens democracy around the world*, Bodley Head, New York.
Reich, R. (2009). *Supercapitalism: The battle for democracy in the age of big business*, Icon Books, New York.
Reeves, R.V. (2018). *Dream Hoarders: How the American upper middle class is leaving everyone else in the dust. Why that is a problem and what to do about it*, Brooking Institution Press, Washington, DC.
Rocca, J-L. (2018). *The making of the Chinese middle class: Small comfort and great expectations*, Palgrave, London.
Rosanvallon, P. (2006). *Democracy past and future*, Colombia University Press, New York.
Rozelle, S. (2020). *Invisible China: How the urban -rural divide threatens China's rise*, University of Chicago Press, Chicago.
Russel, S. & Norvig, P. (2016). *Artificial intelligence: A modern approach*, Pearson, New York.
Sachs, J.D. (2020). *The ages of globalization: Geography, technology and institutions*, Colombia University Press, New York.
Schumpeter, J. (1951). *Theory of economic development,* Harvard University Press, Boston.
Schumpeter, J. (1954). *History of economic analysis*, Oxford University Press, Oxford.
Schumpeter, J. (1989). *Business cycles*, Porcupine Press, New York.
Schumpeter, J. (2002). Seventh chapter of the theory of economic development, translated by U. Backhaus, *Industry and Innovation*, 9: 93–145.

Schumpeter, J. (2006). *Theorie der Wirtschaftlichen Entwicklung*, Dunker & Humblot, Berlin.
Schumpeter, J. (2010). *Capitalism, socialism and democracy*, Routledge, London.
Schwab, K. (2016). *The fourth industrial revolution*, World Economic Forum, Geneva.
Schwab, K. (2018). *Shaping the fourth industrial revolution*, World Economic Forum, Geneva.
Shanahan, M. (2015). *The technological singularity*, MIT Press, Boston.
Simon, H.A. (2019). *The sciences of the artificial: Reissue of the third edition with a new introduction by John Laird*, MIT Press, Boston.
Skarzynski, P. and Gibson, R. (2008). *Innovation to the core*, Harvard Business School Press, Boston.
Skidelsky, R. & Craig, N. (2020). Introduction, in Skidelsky, R. & Craig, N. (eds.), *Work in the future: The automation revolution*, Palgrave, London, pp. 1–7.
Smith, A. (2010). *The wealth of nations*, Capstone, London.
Spulber, D.F. (2014). *The innovative entrepreneur*, Cambridge University Press, Cambridge.
Standing, G. (2014). *The precariat: The new dangerous class*, Bloomsbury Academic, New York.
Standing, G. (2014a). *A precariat charter*, Bloomsbury, London.
Standing, G. (2016). *The corruption of capitalism: Why rentiers thrive and work does not pay*, Biteback Publishing, London.
Standing, G. (2019). *Plunder of the commons, A manifesto for sharing public wealth*, Pelican, New York.
Stilwell; F. (2019). *The political economy of inequality*, Polity, London.
Susskind, D. (2020). *A world without work: Technology, automation and how we should respond*, Allen Lane, London.
Susskind, J. (2018). *Living together in a world transformed by technology*, Oxford University Press, Oxford.
Szeliski, R. (2021). *Computer vision: Algorithms and applications*, Springer, London.
Teece, D.J. (1988). Technological change and the nature of the firm, in Dosi, G.; Freeman, C.; Nelson, R.; Silverberg, G. & Soete, L. (eds.), *Technical change and economic theory*, Pinter, London, pp. 356–381.
Tucker, R.B. (2002). *Driving growth through innovation: How leading firms are transforming their futures*, Berrett-Koehler Publisher, San Francisco.
Unterberg, B. (2013). *Crowdstorm: The future of ideas, innovation, and problem solving is collaboration*, John Willey & Sons, London.
Vangeli, A. (2019). A framework for the study of one belt one road initiative as a medium of principle diffusion, in Xing, L. (ed.), *Mapping China's one belt one road initiative*, Palgrave Macmillan, London, pp. 57–89.
Vogel, E,F. (2011). *Deng Xiaoping and the transformation of China*, Harvard University Press, Cambridge, MA.
Von Hippel, E. (2005). *Democratizing innovation*, MIT Press, Cambridge.
Weeks, J.F. (2014). *Economics of the 1%: How mainstream economics serves the rich, obscure reality and distorts policy*, Anthem Press, New York.
Weick, K.E. (1979). *The social psychology of organizing*, Wiley, New York.
Xing, L.; Giger, M.L. & Min, J.K. (2020). *Artificial intelligence in medicine: Technical basis and clinical applications*, Academic Press, New York.
Zaltman, G., Duncan, R. & Holbeck, J. (1973). *Innovations and organizations*, Wiley, New York.
Ziblatt, D. (2017). *Conservative parties and the birth of democracy*, Cambridge University Press, Cambridge.
Zelinka, I. & Guanrong, C. (2019). *Evolutionary algorithms, swarm dynamics and complex networks*, Springer, London.

5 Competence transformation

Core idea in this chapter

As social systems seek to maximize the potential of automation and artificial intelligence to increase productivity, it is essential for established competence to be transformed in order to ensure that it is compatible with new technologies.

Key points in this chapter

- Creativity and innovation will be important, even decisive, as we attempt to protect ourselves against the impact of robot shock.
- Despite the huge investments being made in robotics, it will take time for productivity to increase. This is because the competence required to utilize robotic automation is not yet available. The consequences will be higher costs and increased automation, but stagnating – even declining – levels of productivity.
- Developing 'Sequential Career Commonality Utilization' can prevent simultaneous increases in productivity and costs.
- Many years of higher education is no longer a way to ensure upward social mobility.
- The more impact that robot shock has on the labour market, the more likely we are to experience restrictions on our individual freedoms.
- For most people, the likely result of robot shock will be a descent down the social ladder into the growing ranks of the precariat. A few people will benefit from robot shock, however. This minority will enjoy increased influence and wealth with a very few of them ascending to the ranks of the elite.
- Competence transformation will hasten the robotization of most processes in the workplace.
- Countries that control the competence and raw materials necessary for the manufacture of semiconductor chips will be the most successful at tackling robot shock.
- As demand grows for higher levels of competence transformation, more stringent performance criteria will most likely be applied across the system as a result.

DOI: 10.4324/9781003567325-5

Introduction

In this chapter, we will examine the following question: What is the potential impact of transformation in competence on future employment trends? When we talk about competence transformation, we mean that people's existing competence will need updating as new technology comes on stream.

It is often said that individuals and social systems must adapt to reflect current market demands. As a starting point, this is correct. It is important to note, however, that our ability to adapt is finite. In the following, we identify seven processual limitations on the ability to adapt.

1. Scope of opportunity: First, we cannot assume that people will be universally successful in adapting to market demands.
2. First-mover advantage: Second, clearly some people will be quicker than others to understand how they need to adapt. People who adapt early will have first-mover advantage,[1] unlike those who adapt later.
3. Education: Third, it is always difficult to know what requirements the market will impose on the kinds of competence that will be in demand in the future. The future we are talking about here is the one to which current students will need to adapt.
4. The consequences of globalization: Fourth, globalization – regardless of future developments – will influence demand in the labour market and make it even harder for both individuals and social systems to identify the best ways to adapt.
5. Limitations on resources: Fifth, all education requires resource inputs from both individuals and society. We can't assume that all talented individuals will have the economic resources necessary to embark on education programmes that will ensure them secure, well-paid jobs in the future. If a person does not possess the kind of education that the market demands, then that person may find themselves in a situation where they can neither adapt to current labour market requirements nor find a job in the future.
6. Lack of social mobility: Sixth, people who have neither the skills, talents nor economic resources to access the necessary education and training will find that their options for social mobility are limited.
7. Meritocracy: Seventh, as a meritocracy develops, people will pull up the ladder behind them, making it more difficult for others to join their ranks. Their motive will be to keep their own wages high.

We see that there are many restrictions on people's opportunities to adapt to the demands of the market and the outside world. Once the ability to adapt is restricted, freedom of choice and personal freedom will also be curtailed.

The robot shock, involving the applications of AI and robotics, will result in limitations to people's ability to adapt. In other words, many people will not have the sufficient competence to deal with the new technology. It is not just a matter of adapting to new technology; the reality is that this will not be possible for many, as has been the case in previous industrial revolutions.

There are two aspects that are important to understand in any industrial revolution. First, there is a time lag, and second, there are the restrictions mentioned above that limit the individual's freedom.

Time lag simply means that it takes a certain amount of time from when an action or event occurs until people and social systems are able to react. The time lag is probably the most important factor in change processes. When new innovations create gaps in the market, this creates an imbalance. This is the time when all known rules and procedures come into play. When entrepreneurs enter the market to fill these gaps, they acquire the expertise they need from the market. This sends signals to people that they need to develop their competence. Therefore, many people will attempt to acquire new skills that are compatible with the new technology. After a while, the entrepreneurs will have filled the gaps in the market that the innovations created, and the market will again be in balance. In other words, the time lag is the time it takes from when the gaps in the market are created by the innovations, until they are filled by the innovative entrepreneurs. It is this time lag that determines the overall meta-structure, that is, how people and social systems manage to adapt to the new technology.

There is another element in the limitations concerning how people can adapt to new technology. This element can be formulated as a proposition: The more competence that exists in the population in relation to the market's demands, the greater the probability that job security will be less, wages will be lower, and the demand for a worker's competence will also be lower. In other words, price is dependent on the interaction between demand and supply in a market. If a social system develops a surplus of skills that are in demand, this will push down wages, and push profits up. This occurred during the period of globalization from the 1970s to the 2020s, when there was a surplus of labour (Gerstle, 2022). Many workers in the US became frustrated and resentful (Frank, 2004); many felt powerless (Frank, 2020) as well as alienated (Carney, 2020). It was difficult to find a job, and workers became 'nomads' wandering around the country in search of work (Bruder, 2017).

When the intelligent robots occupy the natural habitat in which people have lived for many generations, the conditions for survival in this habitat change. The requirements for competence change; the demand for the old competence decreases, and the opportunities for gaining new competence in the education system increase. All these factors will change the norms and values that previously existed in relation to the labour market. In such a situation, many people will feel that they have been deprived of the security which they have built up throughout their lives. This new complexity will suit some people more than others. In this process, some people will fall down the economic and social ladder, while many will be left gasping for air in the struggle to survive. However, a few will manage to adapt and climb up the social ladder.

It is only the few who will manage to change and develop their skills in line with the demands of the new technology. When a certain amount of time has passed, winners and losers in the labour market will become visible. It is during this time lag that the technology will have become part of the standard technology that operates in social systems. This does not mean that everyone else will become unemployed or a nomad in their own country (Bruder, 2017). What we emphasize here is that after the dust has settled and the time lag has passed, there will be some clear tendencies of the robot shock.

1. The first tendency will be a greater degree of economic and social inequality.[2]
2. The second tendency will be that the vast majority of people will be employed, but many jobs will be insecure, poorly paid and in many cases based on unsatisfactory short-term contracts (Standing, 2014, 2014a, 2016, 2019; Johannessen, 2020).

Those who do well after the shock of change brought about by the introduction of robots in working life will be able to climb up the social ladder, but they will pull up the ladder behind them, making it more difficult for others to join their ranks. This can be explained in relation to a social survival instinct, or the principle of self-interest, which Adam Smith described in his book, *The Wealth of Nations* (1776);[3] the book emphasized the importance of people's self-interest for economic growth. Self-interest is viewed as one of the basic laws of capitalism. This economic ideology has been elaborated on by many others, such as Paul Samuelson, who has been called the 'Father of Modern Economics'. He is the author of the influential textbook, *Economics*, which is perhaps one of the most popular economics textbooks of recent decades.[4]

There are two types of people and social systems that will be able to survive the robot shock. One type is those who manage to create their own future. The other type is those who manage to adapt to what others have created. So it is not only those who manage to adapt who will be able to survive the robot shock. It will also be those who are creative and who are able to generate innovations. What we are saying here is that creativity and innovation will be important, even decisive, as we attempt to create our own future at a time when the robot shock will create waves far beyond the technological field.

Of course, it is not the case that those who are unable to create the new, or those who are unable to adapt, will disappear or die out. On the contrary, these people will both survive the robot shock and be in demand in the future labour market. However, their wages will be much lower than today. In other words, they will be in demand because the robots will lack the human touch that is needed in some sectors, such as the health sector. Thus, there will still be a need for workers with 'caring hands' that can take care of the elderly, the sick, the disabled and so on. However, having a job in the future labour market will not be the same as having an income you can live well on; it will be an income you can survive on, and that is something quite different. This development, where wages for workers stagnated, and also declined, both relatively and in real terms, started during the neoliberal period from the end of the 1970s (Gerstle, 2022; Petras & Veltmeyr, 2011).

In addition to the seven procedural limits to adaptation mentioned above, there are three crucial structural limitations that are of fundamental importance.

1. First, humanity cannot adapt to temperature rises that are harmful to its own and the world's biological system. Therefore, the fight against climate change is crucial for humanity's survival.
2. Second, the global food supply is diminishing related to a growing global population.
3. Third concerns the oxygen level in the air we breathe.

These crucial issues are not only necessary prerequisites if we are to survive the robot shock, they are crucial to the future of humanity. Therefore, competence that can generate innovations that are related to these issues will be important in the future labour market. This applies to everything from the reduction of CO2 to the development of food production from areas we have not previously thought of as food for humans. Thus, the people in demand will mostly possess STEM skills.[5]

Of course, before we investigate the workplace of the future and the possible 'robot shock', we have to deal with the crucial issues regarding the survival of humanity mentioned above. In other words, first things first. When we have dealt with the crucial issues, we can then reflect on the seven limits of adaptation mentioned above. We are aware that there may be more than seven such limitations. Therefore, everyone who has input into the debate about the limitations of adaptation, and possibly other crucial issues, should contribute to this knowledge so that we can know with greater certainty which competence transformations will be necessary to cope in the future labour market.

When people and social systems have to adapt to changes in the world around them, the first thing that must be done is to examine what limitations to adaptation people and social systems have. The seven processual limitations on the ability to adapt mentioned above will need to be taken into account when people and social systems carry out the necessary changes. If this is not done, people and social systems will be exposed to the robot shock without being able to do anything to dampen the shock.

The robot shock is not just something that will be experienced as annoying and troublesome. It will affect the individual's and businesses' opportunities to survive in working life. If they are unable to cope with the new competence requirements, the individual will become part of the growing precariat (Standing, 2014, 2014a, 2019) and only be able to find poorly paid insecure work.

The robot shock can be related to the feelings of powerlessness and alienation an individual may experience due to not having the competence that is in demand; in other words, there will be a distinction between those people who have competence that is in demand and those who do not. This factor will say something about the degree of robot shock the individual will be exposed to. Different people and different occupational groups will be affected differently by this competence distinction. There are many indications that those who hold positions where decisions can be made by intelligent robots using evolutionary and learning algorithms will be the ones most exposed to the negative aspects of this competence distinction. The rationale is that it will be possible to use algorithms to carry out work functions and decisions based on logical intelligence. For example, this decision-making may concern decisions that will benefit businesses, such as strategic decisions for future planning. In other words, it will be possible to use AI and robotics' applications to deal with those work aspects of a business concerned with figures, quantities and calculations.

The robot shock in relation to competence distinction will affect people in many ways. Some people will express opposition to changes. In any change process, people will often be anxious about the future or express hostility to anything new that can damage their current position. However, others will do what they can to adapt to change and improve their situation.

Possibly, the robot shock will lead to many people becoming apathetic, because they will feel that they are unable to relate to the emerging complexity caused by the introduction of new technology. Some people will become anxious about the fact that the emerging complexity can lead to a loss of control regarding the development of new technology. For instance, Stephen Hawking, and the other signatories, who signed an open letter expressing concerns about the future development of AI.[6] Moreover, it seems probable that many people will become passive regarding new technological developments. They will very likely develop action paralysis and a form of apathy towards technological development. What we can also very likely experience will be waves of innovations associated with the new technology related to AI. This will be experienced as unpredictable changes by people and social systems. These waves of innovations will also result in people and social systems reacting in unpredictable ways. If the pace of change outstrips people's ability to adapt, forms of erratic behaviour will emerge at the individual, organizational and societal levels. Erratic behaviour leads to a loss of commitment, motivation, energy and the ability to make the necessary adjustments, which Prospect Theory has illuminated (Kahneman, 2011; Kahneman & Tversky, 1979, 2000).

The robot shock related to competence distinction can lead to at least four consequences:

1. Erratic behaviour resulting in a loss of resources.
2. People will attempt to protect themselves from what they perceive as hostile developments, resulting in diminished social cohesion.
3. Action paralysis, resulting in reduced commitment and motivation.
4. Anxiety, affecting the individual's mental state and his/her performance and the performance of the social system.

In relation to the above description, competence transformation can be related to the resistance to change and to some of the negative consequences of the robot shock, which we can call the 'robot syndrome'. These consequences are internally linked. The internal links in this robot syndrome will lead to the fact that if the limits of one variable are exceeded, the entire robot syndrome will easily result in pathological reactions. The interconnections will lead to increasing complexity and difficulty in coping with these interconnected symptoms. The paradox may be that this syndrome can only be tackled by using the technology that initially led to this development. In other words, artificial intelligence and intelligent robots will first promote a complexity that triggers a social syndrome. Then it will be the same technology that will be needed to solve and cope with the syndrome. The obvious consequence for the people involved in this process will be feelings of alienation, powerlessness and despair.

Narratives related to competence transformation

Case letter 1: The productivity paradox related to artificial intelligence and automation

Today, robotics and AI have become important for what is called robotic process and intelligent automation.[7] This is a completely new way of promoting efficiency, productivity and automation in administrative work functions. Possibly this can be compared to the time when digitization began to have consequences for the production and distribution side of businesses. This development and application of technology to the information processes in the administrative systems of private and public organizations could have major consequences for how we organize and apply this technology.

To a large extent, the development of algorithms and software will enable the further development of the automation processes that are already being implemented in the administrative systems of private and public organizations. An important point related to all AI automation concerns the competence of those people who implement the automation processes. If these automation processes seem complex and impenetrable to these people, then the perceived risk when implementing the robotic process and intelligent automation in systems will increase. When the risk is felt to increase, people will develop strategies to try and reduce this risk. One of these strategies will be to limit the use of this technology, because the consequences will be difficult to foresee. Another strategy will be to use the automation tools, but at the same time make use of the old working methods. In both of these examples, costs will rise, productivity will fall, even although automation increases. In this context, it is perhaps appropriate to rephrase a statement made by the Nobel Prize winner in economics, Robert Solow: 'We see automation processes everywhere, just not in the productivity figures'.[8]

Solow's statement also refers to what is called the 'productivity paradox'. There are many explanations for the limited increase in productivity in American industry in the 1970s and 1980s, despite large investments. One of the explanations that may be of interest when discussing the current new wave of AI automation is what we term here 'competence transformation', that is, it takes time before the old competencies are transformed so they are compatible with the new technology. During this period of time, investments will be made in AI automation processes, but productivity will not increase comparably. In other words, a relationship will emerge regarding an increased degree of investment in AI automation and a falling rate of productivity. If this occurs, it is conceivable that the development of automation processes will be hindered for a time, because it will be difficult to argue for increased automation when there is no resulting increase in productivity. This is by no means a new insight. Erik Brynjolfson reached the same conclusion regarding the productivity paradox in the 1970s and 1980s in the US.[9] What is new is the link between Solow's statement, Brynjolfsen's findings and the comparison with AI automation processes being further developed today.

Although there are many who do not agree that there was a productivity paradox in the American economy in the 1970s and 1980s related to information technology, the vast majority agree that there is a learning perspective that should be taken into account when new technology enters established work processes.[10] In other words, it is important to link the learning perspective to the introduction of new technology in order to increase productivity optimally.

The proposition that can be derived from the reflections in this case letter is as follows: As social systems seek to maximize the potential of automation and artificial intelligence to increase productivity, it is essential for established competence to be transformed in order to ensure that it is compatible with new technologies.

Case letter 2: Managerial competence, automation and decision errors

An ever-increasing number of work functions are being taken over by automated computer processes. These processes were initially aimed at logical, instrumental and routine-based decision-making processes. As artificial intelligence has been further developed, the technology is now also able to handle decisions that require cognitive assessments and reflections in various situations; in other words, most work functions have now been affected by artificial intelligence.

When AI technology is used in decision-making to an ever-increasing extent, leaders need to be familiar with the technology. However, it is often the case today that very few leaders have extensive knowledge of the AI technology used in decision-making processes. This situation results in many leaders feeling powerless and alienated in relation to their inability to explain how decisions are made using the technology. Their alienation to the technology also affects the relationship with the people they are in charge of. If the leader is unable to explain why certain decisions have been made, then it negatively affects his/her authority.

Automated decision-making involves the use of data, AI machines and algorithms to make decisions in work processes in the public and private sectors. In other words, leaders should have knowledge of algorithm design, amongst other things, so they have oversight of the decision-making processes, and so they can intervene if need be.

Zuboff (2019), Li (2020), Ford (2016, 2021) and others have shown many examples where the use of algorithms in everyday situations can lead to negative consequences for customers and where those people who operate the decision-making systems or their leaders are unable to explain or alter how decisions are made. For example, in the case of applying for a bank loan, applications have been rejected because the applicant had an 'unsatisfactory' CV with 'gaps' in it due to a longer period of illness. In some cases, the loan applications were successfully re-assessed, while in other cases, they were not, because those responsible for processing the loan were unable to intervene and change the decision due to lack of competence in overriding algorithms.

It is of course nothing new that decisions lead to mistakes and problems arise. However, when people are responsible for the decisions, it is possible to reflect on what went wrong and then change and correct the decisions. Of course, there may be psychological factors that make a leader unwilling to admit that a mistake has been made. This is a situation that people recognize and which it is possible to do something about. For instance, amongst other things, you can go to the level above the leader in question and take up the problem with the leader's superior. When it comes to decisions made by artificial intelligence (AI), the picture is not as clear. Of course, the new technology does not have psychological limitations that prevent the correcting of obvious errors. The decisions made by AI applications are often based on algorithms that have been designed by an algorithm engineer. However, it is improbable that the engineer would be readily available for consultation, regarding the numerous decisions made by the AI application, as these engineers are often located elsewhere around the globe and have numerous other tasks to deal with. Thus, any errors that are made by the AI application need to be addressed by a leader or a team of experts in the organization where AI is utilized for decision-making. In other words, this concerns productivity improvement and quality improvement related to the AI automation processes. This will require that the leader and team have expertise in AI technology so they are able to deal with any problems that arise. If the leader and the team do not have the necessary competence, then although many incorrect decisions are made, no one will have the competence to do anything about it. One of the reasons for the many errors that have been reported in relation to decisions made by AI applications is that many algorithms are not designed in such a way that they include all the necessary information about the people they make decisions about. For example, information related to factors such as illness, ethnicity, sexual orientation and so on. In other words, so that automated decisions are not based on incomplete information, they need to be designed so that they take into account differences in the people they are making the decisions about. If the algorithms do not have this capacity built into their structures, then the decisions will be both wrong and at the same time difficult to correct. This has been shown to be the case in loan applications, as mentioned above. We have also seen decisions in medical diagnosis and treatment that have been assessed incorrectly, because the automated process did not take into account normal differences that can occur in people from different backgrounds.

One of the solutions to ensure that AI applications do not make incorrect decisions is to have a parallel decision-making apparatus controlled by people so as to ensure both fairness and equality in the decisions. However, such a parallel decision-making apparatus would increase costs and reduce productivity. The more long-term solution would be to raise the competence of the leaders and operators who stand between the decisions made by AI applications and those who are affected by the decisions. However, this transformation in competence takes time. There is therefore much evidence that some organizations will maintain parallel decision-making apparatuses, while developing the competence of those people in the organization who process the AI decisions. As remarked above, this will lead to increased costs. Competence transformation affects not only decisions but also the cost structure of organizations.

Description related to competence transformation

In the past, education has been the key to social mobility for the working class and the middle class. The educational studies that have provided the basis for finding jobs with high income and status have been closed to most people but open to those who achieve the very best results in upper secondary education. Such educational studies have enabled graduates to secure a place high up in the economic and social hierarchy. This concerns professions, such as doctors, engineers, lawyers, economists and so on. During the post-war period, many years of higher education ensured that graduates could find well-paid, secure jobs. This situation has not changed that much in the innovation economy. However, it has become more advantageous to complete an education that is compatible with the new emerging technology, such as AI, robotics, Big Data analysis, algorithm design and so on. Graduates who have completed this type of education can often hope to move up the social ladder and join the upper middle class. Globalization from the 1980s onwards, when China opened up to Western capital, has strengthened and increased the importance of higher education as a social mechanism for securing a position in the upper middle class.

However, globalization has added a new element to social mobility. This concerns the importance of entrepreneurs and their innovative power. The rise of entrepreneurship as a driving factor for social mobility has also lessened the importance of traditional extensive higher education as a means of securing a future in the middle class (Ainley, 2014, 2016). Sometime around 2020, a long university education was no longer as important for future-proofing one's own position in the middle class, as it had been between 1945–2020. However, young people today are still told that they need to complete a long university education in order to secure their future in the labour market. In this context, Ainley's book, *Betraying a Generation: How Education is Failing Young People* (2016), makes a valid point; that is, higher education no longer guarantees that graduates will be able to find secure and well-paid jobs.

The labour market still demands people with up-to-date educations. Today, this concerns expertise in the use and development of the new technology, especially all the technology related to artificial intelligence; the demand for this type of expertise complicates the future-proofing of people's qualifications. For instance, the psychologist will need additional skills to increase his/her productivity. This additional education concerns technological competence related to the practical application of artificial intelligence. For instance, the psychologist will need to have cutting-edge expertise in coding and algorithm development. The psychologist who has this additional competence will be able to increase his productivity by at least 100 per cent. The idea behind this very high productivity growth is justified by the fact that if the psychologist did not have this competence, then the competence would have to be purchased. This type of competence is in great demand, and therefore the price is very high.

Globalization affects the demand for the additional competence that we have discussed above, the 'dual competence' that the psychologist in our example needs to have in order to be more productive than his colleagues. In concrete terms, many specialized engineers are trained in the new technology in China. These engineers can compete in a global market, because technology can flow across borders at the touch of a key. This development affects the labour costs of businesses that want to purchase expertise in a market.

When China opened up to Western capital, industrial workers in the West faced competition due to the low labour costs in China. This resulted in the outsourcing of work to China. China was eventually able to trade its way to prosperity. Amongst other things, China used this wealth to invest in mass higher education. In addition, the Chinese invested in engineering education with specialization in all the technology related to artificial intelligence (Jain, 2021). An important point here is that the Western knowledge workers who work with AI technology are meeting competition from the Chinese AI engineers. The Chinese engineers can sell their expertise and the products they produce at a significantly lower price than the Western AI engineers. This may result in a situation similar to the one we described above concerning the loss of industry jobs to China; that is, Western AI work may be outsourced to China due to lower labour costs.

In this context, we can relate the demands of the market to the consequences of globalization and the importance of higher education for future competition within the field of high technology. In order to reduce high-tech competition from China, the Biden administration took action in the autumn of 2022 and introduced new restrictive legislation. The US could not openly state that this was done to give them an advantage in the competition with China in the field of AI technology, because this would have contravened the World Trade Organization's (WTO's) regulations. Therefore, the US argued that the legislation was being passed due to national security issues; that is, that the new technology developed in China and sold on the American market was a threat to US national security. We will probably never be able to ascertain the truth of this 'national security argument'. However, the US Chips and Science Act of 2022[11] hindered the free competition in high technology between China and the US and can thus be claimed to be a type of protectionism. Moreover, the US has managed to enrol the support of other countries in the 'technology alliance' against China.

Economic history informs us that the nation that leads in technology will also be able to dominate economically and militarily (Frey, 2019, 2020). This is why it was so important for the US to take action regarding the development of high technology in the innovation economy. It must be said that China did not just wait and see regarding the US's negative reactions concerning China's technological development. From the first signals that the US could no longer withstand the competition related to Chinese low-cost labour and high-tech development, China has implemented its own measures to counteract US initiatives. The first signal of the US's opposition to China's growing power became evident when the former Secretary of State, Hillary Clinton, published her 'America's Pacific Century' in the *Foreign Policy* journal in 2011. Tensions increased in strength under President Trump when the US launched a trade war against China. Tensions further increased during the Biden administration with the statements of the Secretary of State, Antony Blinken, who, in 2021, accused China of committing genocide against the Uighurs in China's Xinjiang province.

China's counter-reaction against the US's aggressive stance consisted of three strategies:

1. First, from 2012 onwards, just after the publication of Clinton's article, China launched its New Silk Road, a vast network, which included more than 100 countries, stretching across Eurasia and other continents with the aim of stimulating economic growth and trade (Johannessen, 2021, 2023a, 2023b).
2. Second, China launched its 'dual circulation' development paradigm in order to sustain growth and develop a strong domestic market, amongst other things. The aim is to meet the growing demand from a growing Chinese middle class, which comprises approximately 700 million people.
3. Third, China joined the RCEP, the world's largest free trade agreement (1 January 2022).

These three strategies can be understood as China's plan to meet the American technological competition. Of course, it is not just technology competition that is at stake between China and the US; the competition also has the potential to trigger a conflict between the US and China.

In the past, education has, as indicated above, been the social mechanism that has promoted social mobility. If an individual failed to complete a good education, then his/her whole future was at risk. However, people's insecure futures could not be placed at the door of governments or those people who steer social development. At least, not when one of the dominant prevailing ideas in society is that people are responsible for forging their own path to success. In other words, this popular view expresses that it is the responsibility of the individual to make sure he/she completes a good higher education so they will be able to find a well-paying job to secure their future. The idea is that people are solely responsible for their educational success or failure and that this is not a social responsibility. However, this proposition hardly holds true in those countries where it is mainly the role of parental wealth that determines whether or not a young person can complete a good higher education. Of course, there are exceptions; most countries also provide scholarships, grants and financial support to enable less well-off students to gain access to university studies. This whole process of education as a social mechanism for social mobility promotes development towards a meritocracy.[12] In such a system, power is distributed to those who have achieved specific competences. The opposite of the meritocracy is, amongst other things, those systems where power is inherited, such as in the Middle Ages, or those systems where power is transferred through generations to the families that own capital. The importance of meritocracy has been crucial to the growing middle class in both the West and

China. It is in this way that they have been able to develop and consolidate their position in society.

Competence transformation can be understood in relation to the five elements reviewed in this description. These five elements are: The demands of the market; globalization and the resentment sparked by grievances against globalization; education; and social mobility, which is largely secured through education; and, meritocracy, which has been reinforced by market demand and globalization. Underlying these five elements of competence transformation, however, lies the new technology's demand for competence. To simplify this, the new technology can be understood as artificial intelligence. We have also mentioned intelligent robots, Big Data, coding and algorithm design. The point is that it will be artificial intelligence that will constitute the general purpose technology (GPT), which will change most, if not all, work processes in the working life of the future.[13] In other words, the transformation in competence will affect the development of the emerging technology and vice versa. There is a systemic connection between artificial intelligence, intelligent robots, competence transformation and the workplace of the future. Thus, competence transformation affects the development of the workplace of the future via links to the emerging technology. In its consequences, this means that technological education will become more important in relation to upward social mobility. This also means that many other types of education, which do not include a focus on the new AI technology and related technologies will be undervalued.

Based on the above description, we have developed a conceptual model that indicates some elements of competence transformation in the innovation economy (Figure 5.1).

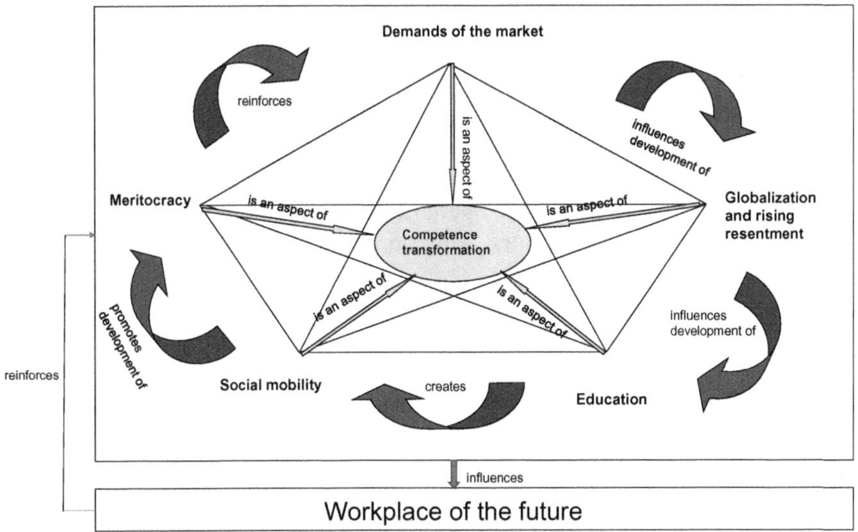

Figure 5.1 Competence transformation: A conceptual model.

In the following, we will elaborate on the description of the five elements in Figure 5.1 in relation to competence transformation.

Demands of the market

The market's demands can be many. One such demand is the need to adapt to change and to renew skills continuously. One needs to adapt to a rapidly changing world, when the pace of technological change is greater than ever. However, what may be an advantage in some contexts can become negative when the changes never seem to end. Then precisely what was a strength, the fact that you are willing to change, can quickly become a weakness. When the positive becomes a negative factor, you will have exceeded a limit. It is when this limit is exceeded that adaptation no longer works positively as a change mechanism for the individual. This adaptation applies especially to the continuous renewal of skills.

The market's demands for new skills and the development of globalization are very closely linked. The neoliberalism associated with the former US President, Ronald Reagan and the former UK Prime Minister, Margaret Thatcher (Gerstle, 2022) facilitated foreign investment in Deng Xiaoping's China in the late 1970s and beyond in the 1980s (Vogel, 2011; Deng, 1994). It was precisely this interaction between China and the West that made it possible for China, over a 30-year period, to develop mass competence, where the focus was largely on developing China's soft power, i.e., China's technological competence (Jain, 2021). As mentioned, China became so adept at adapting to the market's demands related to technological competence that the US felt compelled to adopt a law, the Chips and Science Act (2022),[14] which would prevent the US from promoting technological competence in China. With this development from 1980 to 2022, one can say that China became so good at adapting to the demands of the market, regarding new skills, that it led to a conflict with the US. The increased tensions between the two countries could possibly lead to a future military conflict (Allison, 2018). This shows that we can be so good at adaptation, both at individual level and at the system level, that it can lead to us creating conflicts and economic setbacks, where we wanted to create progress and economic growth.

The Chips and Science Act will provide $50 billion for American semiconductor research and development. The purpose is quite clear. The US aims to bolster US leadership in semiconductors, while hindering Chinese development in this industry.

Neoliberalism is a free-market economic ideology that promotes the deregulation of markets and industries. This ideology became the guiding star for economic policymaking in most Western countries. Neoliberal policies led to the rich getting richer, while greatly increasing income inequality (Weeks, 2014; Dorling, 2019). During the period when neoliberal policies were adopted globally, during the late 20th century and the 2000s, China's economy grew very strongly, but there was also increasing income inequality.[15] The Chips and Science Act (2022) may be viewed as the result of a growing resistance in the US to the increase in Chinese soft power. China's technological development, and thus its increase in soft power, is closely linked to the development and transformation of its educational system. As mentioned, the first signal of the US's opposition to China's growing soft power became evident when the former Secretary of State, Hillary Clinton, published her 'America's Pacific Century' in the *Foreign Policy* journal in 2011. Since 2011, tensions have further increased quite considerably between China and the US.

The free market focuses on the unrestricted competition between businesses, but it is not so much focused on development of expertise (Sandel, 2020: 62). This can perhaps partly explain why the US did not overly focus on technological development in recent years, which they now have done with the Chips and Science Act. China, on the other hand, rode two horses. They bet on the market on the one hand and invested in the development of skills and expertise, on the other. This development resulted in, as we have shown above, the adoption of the Chips and Science Act, which some critics view as being opposed to free-market philosophy, that is, that government should not interfere in the free-market system.

In the above discussion, we have attempted to show how the market's demands can influence the development of the workplace of the future. On the one hand, we have seen the free market operate largely unhindered without government intervention in the US, from the 1970s up until the Chips and Science Act of 2022. On the other hand, we have seen a balance between the market's demands and government intervention in China and how this balance has influenced the development of skills for future workplaces in China.

Globalization and resentment

Major changes in economic, political, technological and social systems lead to periods of decline in efficiency and productivity.[16] After a while, however, productivity increases, mainly because technology and technological competence have become part of everyday working life for most businesses. However, some businesses and people will be unable to make the transition from the old working methods to the new methods. Businesses that try to hold on to the old methods for too long will not be able to compete in the market. Globalization from the 1970s until the 2020s showed this development quite clearly. Many workers in countries such as the US were made redundant when work was outsourced to China, where wages were low. Thus, American workers were unable to compete with the Chinese workers due to low labour costs in China. Globalization hugely increased foreign investment in China, especially by American investors who made huge profits, while many American workers lost their jobs. In addition, globalization resulted in an erosion of the US middle class (Reeves, 2018). Thus, globalization negatively affected both the American industrial workers and the American middle class, leading to feelings of resentment amongst large sections of the population. This situation resulted in a politics of resentment which provided fertile ground for the rise of populist sentiments, more specifically the MAGA movement ('Make America Great Again') led by Donald Trump. Politicians said outright that China had stolen American jobs. The fact that it was American capital invested in China that led to the outsourcing of work to China was conveniently swept under the carpet. Resentment against the negative consequences of globalization was possibly one of the reasons why Trump won the presidential election in 2016. After Trump lost the presidential election in 2020 and Jo Biden became president on 21 January 2021, the criticism of China intensified and took on new acerbic tones, when the US Secretary of State, Antony Blinken, accused China of committing genocide against the Uighurs in China's Xinjiang province.

De-globalization began around 2022 and was intensified by the Russian invasion of Ukraine. In addition, the US Chips and Science Act, adopted 9 August 2022, included export restrictions aimed at hindering China's development of its high technology.[17]

If we look at China's development from the late 1970s to the 2020s, the following developments will help explain how globalization has affected Chinese workplaces.

1. First, China was able to produce goods at a low cost due to low labour costs and was thus able to trade its way to prosperity. On the other hand, wages increased in urban areas in China, while rural areas did not see the same increase, resulting in income inequality between rural and urban areas (Fan et al., 2009).
2. Second, this inequality was an intentional governmental policy implemented by the Chinese leaders. Xiaoping stated quite explicitly that some people had to become rich before others (Deng, 1994). The idea was that when some became rich, they would help others climb up the economic ladder (Vogel, 2011).
3. Third, the 'Chinese Dream' is to raise everyone in China up to a middle-class income through collective effort (Brown, 2018).
4. Fourth, globalization and China's economic growth made it possible to expand to other regions and countries. The New Silk Road (also called the Belt and Road Initiatives [BRI]) was implemented in 2012 by the Chinese leader, Xi Jinping (Batzhan et al., 2020). Around 2020, there were approximately 130 countries and institutions that had signed various project agreements with the Chinese government (Beeson & Li, 2016; Berlie & Hung, 2020). The Belt and Road Initiatives promoted the development of jobs in China and in those regions where the Chinese projects were developed, e.g., in Pakistan and many countries in Africa (Drache et al., 2019; Doshi, 2021).
5. Fifth, China used much of its income from the export of cheap manufactured goods to the West to invest in its own education system. This applied to both mass education and higher education (Jain, 2021). Chinese students also studied abroad in large numbers at universities in the US, the UK, France, Russia and other countries. China also developed a large number of its own universities. This investment in soft power eventually resulted in China becoming a major player in high-tech development. The development of high-tech skills at Chinese universities meant that Chinese businesses were able to utilize these skills in their production. This development of high-tech skills also strengthened the growing Chinese middle class (Li, 2021).

6. Sixth: A focus on the development of AI technology. Chinese President Xi Jinping has emphasized the importance of developing China's capability in artificial intelligence[18] (Johannessen, 2021, 2023, 2023a).
7. Seventh: China's membership in the world's largest free trade agreement (RCEP) will greatly affect future workplaces in China, Asia and Western countries.[19]
8. Eighth: China's implementation of its 'dual circulation' development paradigm will develop a strong domestic market, which will meet the increasing demand from a growing Chinese middle class.

The above description explains how globalization and resentment have affected the development of workplaces both in the West and in China. Around 2022, this development led to what can be called de-globalization (Johannessen, 2023b). This de-globalization has led to increased tensions between the US and China (Allison, 2018). These tensions have manifested themselves in various ways, for example, by the US's adoption of the Chips and Science Act (August 2022).

Education

Education functions as a social sorting machine which has implications for economic equality and inequality. In other words, education contributes to constructing social reality; for various reasons, some young people are given the opportunity to complete higher education, while others are not. Although people in a region or country may have different educational opportunities, they nevertheless belong to a regional or national culture. However, although people may belong to the same culture, education functions as a social mechanism creating distinctions in relation to many different variables.

In earlier times, those people with power and influence could favour relatives and friends by giving them jobs. This type of nepotism was widespread in working life, both in the private and public sectors. In other words, those young people with influential friends or families were able to secure their futures. Today things are different. Young people's futures are to a great extent dependent on what type of education they have. However, those young people with wealthy parents will often have better access to education and careers compared to those young people with poor parents, even when they are talented and have ability. In this context, one might ask: Is this not just a different type of nepotism? In other words, there are educational benefits of being born to wealthy parents; in this context, it might be said that generational social advantage is passed down. In the past, nepotism concerned granting advantages, privileges and positions to relatives and friends in a profession or field. Today, we might say that a new type of nepotism has emerged based on wealth and power. Thus, those young people who belong to a rich and powerful family network will often have much better educational opportunities than those who belong to a poor family network. There are of course exceptions; many countries provide scholarships and grants to enable less well-off students with ability to gain access to university studies. Moreover, some countries provide free higher education, although this is not the case in most countries around the world. In addition, in those countries where higher education is free, parental wealth and education is also an important factor with regard to a young person's educational and career opportunities (Iversen & Soskice, 2019; Przeworski, 2019).

In modern society, in countries and regions such as China, Japan, South Korea, Europe, the US and other countries and regions around the world, a meritocracy has emerged that dominates the development of social systems. This meritocracy is characterized by two strong social mechanisms. First, awarding advantage to people with 'merit', that is, highly educated people with special skills are given the opportunity to reach the highest social level. The second social mechanism relates to what extent a person belongs to a powerful, resourceful and influential network. The connection between these two social mechanisms gives the individual the greatest probability of achieving a position in the meritocracy, which means that this person will be able to secure his/her future and his/her family's future.

There is a systemic link between education and future jobs. On the one hand, education affects the development of future workplaces. On the other hand, future workplaces demand certain skills. In this way, it can be said that the future, through a mechanism of expectations about what will be the most secure and best-paid jobs, affects what type of education a person will pursue.

Based on what has been discussed above, it is quite clear that the educational system functions as a sorting machine for an individual's future position in working life. When a person achieves a good position in the meritocracy, it is also probable that his/her children will also be able to enter the meritocracy and thus end up in a position where their future is also secured. It is erroneous to imagine that educational performance is only dependent on a person's ability. Even if it were the case that an individual with excellent abilities will always be able to manage to complete an excellent education, there are nevertheless questions that can be raised about the meritocratic system. What about those who do not have the ability to acquire the necessary skills that are in demand in order to secure their future? Would these people be better off living in a nepotism society, which could take the responsibility securing their future despite their lack of ability? In this context, it might be said that the meritocracy is a curse for those who do not have the abilities demanded by the new era. In earlier times, the sorting machine was related to who you were related to. Today, the sorting machine is related to one's own abilities. Is it the case that it is more ethically correct that we emphasize an individual's abilities rather than an individual's familial relationships? It is in relation to these questions that one can reflect on the title of Sandel's book, *The Tyranny of Merit: What's Become of the Common Good?* (2020).

Social mobility

Social mobility can lead to success or failure depending on how an individual's education determines which position he/she will assume in the social hierarchy. Since 1945, education has been a key factor in promoting upward social mobility. However, if the findings of Ainley's research are valid (2014, 2016), then, in the future, education will no longer enhance social mobility. If education is not the most important social mechanism for promoting social mobility, then what has become more important? It may appear that who you know has become just as important as what you know; in other words, social relations have once again become an important factor for climbing up the social hierarchy (Chetty et al., 2017; Chomsky, 2016, 2016a). Thus, it seems that the new AI and robotics technology is facilitating two important factors for social mobility. First, there is the importance of cutting-edge expertise that is compatible with the new technology. Second, there is a greater emphasis on nepotism, i.e., who we know matters just as much as what we know, that is, what type of education we have. The necessary prerequisite for this nepotism system to function, however, is that the individual has the expertise that the new technology demands. One of the consequences of this is that a student who pursues a course of studies on the basis of personal interests will not necessarily be able to find a secure and well-paid job after graduation. Upward social mobility will thus involve two factors: who you know and what you know. The individual will need to have an education that is compatible with the new technology and also know the 'right' people, that is, be part of an influential social network.

The robot shock will have a major effect on education. This means that young people will need to think carefully about their educational choices. If a student chooses a course of studies based on personal interests, such as a course of studies in psychology, then it will be decisive for his/her future success in working life that he/she also learns skills that are compatible with the new technology. For instance, these high-tech skills could involve the design of algorithms and the coding of tasks related to diagnosis and therapeutic programs. The professional who has this type of additional competence will be able to increase their productivity considerably, possibly more than 100 per cent. If the professional does not make it known that he/she possesses this high-tech expertise, then the outside world will perceive that he/she has a work productivity that is unique in relation to others. However, as soon as the outside world becomes aware of the fact that the professional has a high level of productivity because of his/her additional high-tech expertise, then the professional in question will not receive the same level of social esteem due to his/her efficiency. Consequently, the best strategy for the professional will be to acquire additional high-tech expertise, but at the same time not make it publicly known that he/she possesses this expertise. However, although this may benefit the individual professional, it will not necessarily benefit the system as a whole.

An important point about the above description can be expressed by the following: Productivity growth can lead to an increase in costs. This is the opposite of what might be expected, namely, that productivity growth results in cost reduction. In the case of the psychologist's additional expertise and increased productivity described above, this may benefit the individual psychologist but not the system as a whole. This can be illustrated by a short narrative authored by Karl E. Weick (1979: 1–2):

> Farmers have been buying heavier and more advanced machinery to save labour. The heavier machines have caused problems. They pack the soil and

> sometimes harden the subsoil and keep water from penetrating to the plant's roots. The subsoil then must be tilled with an even larger, deeper plough which, of course, requires a more powerful tractor to pull it.

In other words, productivity increased considerably but so did costs.

The above narrative highlights the following issues:

1. First, one should reflect on whether productivity is a suitable goal in an age when completely new technology is entering social systems.
2. Second, one should reflect on whether technological, organizational and management innovations should go hand in hand with the development and implementation of the new robot technology.
3. Third, if one does not reflect on the above, one may end up not only increasing productivity but also increasing costs.

Enthusiasm is almost without exception good for the individual and for social systems. The point, however, is that over-enthusiasm can easily backfire and do more harm than good. Therefore, one needs to be cautious regarding the development and implementation of the new robot technology. It can be helpful if the systems that utilize the new technology are capable of making the necessary changes. If a social system is not ready for change, then the new technology may do more harm than good.

If we look again at the case of the psychologist above, he/she increased his/her productivity by more than 100 per cent, which could possibly also enable personal upward social mobility. If one now assumes that this person was able to develop his/her additional expertise because others helped him/her, how shall we then interpret the person's productivity growth. We can express this in the following way: It is because we do what we do that enables you to do what you do. In plain language, this means that the social system (the community), as a whole, helps the individual increase his/her productivity, so that he/she is able to make extra profit, but the community does not get to share in this extra profit. Developing what Weick calls 'Sequential Career Commonality Utilization' (1979: 2) can prevent simultaneous increases in productivity and costs. One such way of organizing social mobility would be to continuously relate the parts to the whole in the working community. In practice, this can be promoted by organizing the entire working community around circular teams, which constantly keep a focus on what the system is designed to do. Concretely, this means a type of organizing that Stafford Beer calls the viable system model (VSM) (Beer, 1994, 1995).

Meritocracy

A meritocracy is a system or society in which people are able to move into positions of success, power and influence on the basis of their abilities and merit. In practice, this is a system where people are given the opportunity to acquire knowledge and skills through education. Meritocracy as a social system contrasts with societies based on inheritance, power and authority, such as authoritarian systems, where the military has taken power and controls the social system.

Post-war societies up until the 1970s were characterized by free market and egalitarian welfare systems, where it was possible to realize meritocratic ideals to some extent. However, by the 1970s, the adoption of neoliberal ideas resulted in an increased focus on market liberalization with meritocratic tendencies (Gerstle, 2022).

Meritocratic systems utilize education as a social mechanism for social mobility and positioning, dominance and power. In the present day, education is increasingly focusing on specific types of expertise, such as expertise in AI, robotics, Big Data and to some extent technologies for gene-editing. In these new technologies, the development and design of algorithms has become crucial to production and distribution processes, as well as behaviour within social systems. When education begins to focus more on the development and design of algorithms, the coding of intelligent machines and so on, then meritocratic ideas will have taken on a new meaning. One could call such meritocratic systems technological meritocracies. These technological meritocracies ensure the position and power of those persons with high-tech educations in fields such as AI and robotics.

When algorithms control our behaviour, then the robot shock will have begun to have real consequences. This control not only applies to how concrete decisions are reached but will also apply to social relationships. In other words, the robot shock will definitely be linked to the development and design of algorithmic systems and how they are systemically connected. These algorithms will eventually be developed further so that they will be able to learn from experience, not unlike the way humans learn.

When society reaches a certain level of robotization, decisions will largely be made by various types of algorithms. This is not science fiction. The use of robotization processes that utilize algorithms are everywhere around us today. The point is that when these robots become based on evolutionary and learning algorithms, they will take over more and more of the decisions that people used to make. In such a system, the freedom to trade in free markets and the freedom to vote in democratic elections will become increasingly curtailed. The market will be largely governed by algorithmic decisions. The results of democratic elections will be a foregone conclusion, because we will largely be controlled by intelligent algorithms that 'tell' us what is in our best interests. Harari (2022: ix) says that we are enjoying an unprecedented technological bonanza, but at the same time facing philosophical bankruptcy. It is the revolutionary development of technology that constitutes the 'technological bonanza'. It is the lack of self-reflection that means we are heading towards 'philosophical bankruptcy'. In other words, the individual no longer needs to think or reflect on their decisions, because algorithms integrated in intelligent robots do it for him/her. If over time we do not use our capacity for reflection and thinking, then it is possible that this ability will degenerate. If this leads to an increase in our social and emotional skills, then it wouldn't be as bad as you might think. The point is that social relationships and our emotional attachments will also be controlled by algorithmic decisions. It is against this background that we can say that Harari's statement above rings true, i.e., our philosophical reflections about everyday life will be reduced to algorithmic decisions. In this way, the robot shock will have affected us in more ways than just what type of education we choose in order to achieve a position in society with status and influence. We will become wholly governed by algorithms, that is, we will become non-sentient beings without the ability to think or perceive. This may sound somewhat exaggerated. However, just think of what may happen if gene-editing tools are developed further and used on a large scale. How then will the individual relate to their own environment? There is one thing that is beyond doubt. The future development of a technological meritocracy will very likely affect our workplaces. This development will strengthen the influence and positions of those with the right type of high-tech education, while reducing the influence and positions of those without such an education.

Analysis related to competence transformation

The emergence of new technology, such as AI and robotics, will affect the political system at all levels. Politicians can aim to create a society where value creation is distributed according to the ideologies they themselves argue for. If this ideology is aimed at economic equality, then it is this type of development that they can promote. If, on the other hand, the ideology is based on the idea that tax cuts for a rich minority creates many jobs, then the new technology may be used to reinforce this ideology. On the basis of this understanding, the development of the workplaces of the future will be shaped by the interaction between technology and political decision-making.

It is becoming increasingly important for people, organizations and countries to be able to conceptualize the desired future. The rationale is that the future is shaped to a great extent by the lessons we have learnt in the past. The decisions made by policymakers will have an impact on what our workplaces will look like in the future. This in itself is not a new insight. The new technology is creating great opportunities for change that must be seized so that people, organizations and countries can take part in shaping the future. The reason is related to the fact that technology is moving faster than ever before. This is also why, amongst other things, Schwab (2016, 2018) calls this transformation the Fourth Industrial Revolution. Industrial revolutions transform societies radically by introducing new technology, new methods of production and organization, new skills and by transforming education. The difference now is that we are in the eye of the technological storm but unable to envision what the world will look like when the storm has passed. What is clear though is that if we are unable to conceptualize, and thus help create the future we want, then we will have to adapt to a future created by others.

The new technology will not only transform organizations and societies in the West but also affect developments in Asia (Johannessen, 2021, 2023a, 2023b). The transformation that will result from the further implementation of new technology in China will have major consequences for the development of future workplaces in both Europe and the US. In this way, the robot shock is a global event, which will affect our common future. Therefore, it is crucial that we are able to conceptualize the future. Without being able to conceptualize the future, we will not have the opportunity to create the desired future. As has been said several times, if we don't create the future, others will do it for us so we have to adapt as best we can. However, this is a passive and unconstructive strategy.

Conceptual strength comes from a historical insight into similar conditions in the past, as well as a creative understanding of how systems develop throughout history. Creativity generates innovations, just as education generates the conditions for the emergence of new technology. In the same way, historical insight creates an understanding of the historical signals sent to decision-makers. It is these signals that can be conceptualized to enable us to understand aspects of the future that will develop. This development is not deterministic but is created by someone for someone. It is therefore important to understand how historical processes have created the basis for the power and positions that exist today. It is this understanding that can then be conceptualized. In this way, we have the tools to change the future. We thus change the future by understanding the social mechanisms that create this future.

Without a map of the future, we will have to adapt to a world others create. The point is that when we enter the future with no notions of what may await us, we will be surprised by all the new transformations we encounter. With limited energy and resources, it will be more appropriate to create a 'map', even if this 'map' does not match the terrain. This is important because we then have opportunities to change the 'map' before we change the terrain with our decisions. The paradox here of course is that we create a 'map' that then changes the terrain and not the other way around. This is precisely what we must do when we enter a time of major technological changes. The technological storm can be dampened and controlled by conceptualizing the social mechanisms that affect the storm. We create the 'map', which we then use to change the terrain we enter.

Regular maps use symbols to label real-life features in nature and make the maps clearer. We then use such a map to navigate an unfamiliar terrain.

However, in the social sciences, especially concerning the field of major technological innovations, it is more appropriate to reverse this process, i.e., we design a 'map' that we then use to create the 'terrain' of the future. Without such an awareness, we will end up in a position where we only adapt, like a chameleon, to the random events in the world around us. Plan or be planned for is the strength that lies in the ability to conceptualize. Harari (2022: ix) expresses that conceptualization is 'a guidebook to the future'. Such a 'guidebook to the future' creates our future workplaces through an understanding of the competence transformation that the new technology and the new future will demand.

The more we focus on technological development, without political control, the more our societies will develop in a purely technological direction. This does not have to be wrong in any way. However, this may result in the 'rational' technological system eliminating from our societies what it means to be human. This appears, amongst other things, in the dialogue between Mustapha Mond and John the Savage in Huxley's *Brave New World.* One can formulate this dialogue in the following way: If everything other than the rational disappears, then human madness has reached its peak. This is where the new technology has the greatest potential to result in ethical collapse. The new technology is becoming so 'rational' that everything is measured, weighed and numbered. If this 'rational logic' is integrated in intelligent robots that control society, then we risk ethical and emotional shipwreck.

Theoretical reflections related to competence transformation

Since the Second World War, education has been the most important social mechanism for social mobility and has functioned as a 'social safety valve' (Sandel, 2020: 156). It is in the period from 1945 to sometime around the 2000s that one can agree with the title of Ainley's article 'Follow your dreams and attend to universities if possible' (2014). Ainley points out that there is a clear connection between a good higher education and the possibility of social mobility. The skills the student learnt during his/her college education ensured that he/she was able to find a relatively well-paid and secure job so that he/she could join the ranks of the middle class.

However, when neoliberal ideology (Gerstle, 2022) and globalization processes left their mark on the economy, something new started to happen. In the US, many of the industrial jobs and middle-class jobs were moved to low-cost countries, such as China (Johannessen, 2021, 2023). Globalization, which most commentators had advocated so warmly, including the Nobel Prize winner in economics, Joseph Stiglitz (2015, 2019), began to show negative consequences. Krugman, another Nobel laureate in economics, asked questions about what the economists had overlooked when the negative consequences began to show themselves (2019). It was probably not so much what the economists had overlooked but the fact that a new global development emerged due to increased technological competition and heightened geopolitical tensions between the US and China. This gradually led towards a de-globalization process, which can be partly explained by the fact that China was threatening to overtake the US in terms of GDP and technological development (Johannessen, 2023b). The sociologist, Richard Sennett (1998, 2004, 2006, 2008, 2013), was perhaps one of the first commentators to express that something new was happening in the global economy. One of these new features was that it began to become clearer around the beginning of the 2000s that a good higher education no longer necessarily provided an opening for social mobility, as it had done before. Ainley provided ample proof for this new development in his book, *Betraying a Generation: How Education is Failing Young People* (2016). He showed the non-validity of the idea that if young people complete a traditional higher education, they will be able to find a secure and well-paid job. In this context, with a reduction in social mobility, the social classes became more rigid and fixed, something the German sociologist, Hannelore Bublitz (2022), writes about when she notes that the elite reinforce this social closedness by the use of hidden codes.

When people with higher education have difficulty in finding good, well-paid jobs in order to facilitate a degree of social mobility, then this is a sign that we are moving towards feudal structures (Kotkin, 2020) and a type of feudal capitalism (Johannessen, 2023). In this development, it is largely automation processes that have put middle-class jobs at risk (Johannessen, 2020a, 2020b, 2021a).

The development we have witnessed regarding the rise and fall of the middle class (Reeves, 2018) is a sign that society is being transformed into something new. This 'something new' can be given many names, like all societies throughout history that have undergone transformations. We choose to term this new society a feudal capitalist society, because feudal structures are gradually emerging (Kotkin, 2020), resulting in a new type of capitalism.

In the US , before the post-war period and before the development of higher education for the masses that provided education to the working and middle class, it was mainly the upper class that had access to university education. In this context, the three most important and prestigious universities were Harvard, Yale and Princeton (Sandel, 2020: 156).

In the 1950s and 1960s, however, many new universities were established with widespread political and financial support. This provided new social mobility opportunities for the working and middle class. However, around 50 or 60 years later, things began to turn around. Many years of higher education was no longer a way to ensure upward social mobility. Empirical studies (Reeves, 2018) indicate that the middle class is declining. One part is being demoted and often end up in insecure jobs with poor pay, the so-called precariat (Standing, 2014, 2014a, 2016). Another part is rising to join the upper class and become ideologues for this class (Johannessen, 2023). The last part is remaining in the middle class and benefitting from the security this position provides (Reeves, 2018).

We have thus seen a development from when there were more closed social structures before the Second World War to a more open social mobility based on education after the Second World War to a new social closedness, where education is no longer a guarantee of social mobility.

Based on the above narratives, description and theoretical reflections, we have developed a typology for competence transformation related to the workplace of the future (Figure 5.2).

On the basis of the typology, we have developed the following propositions:

Proposition: The greater the degree of education and the greater the degree of social mobility that exists in a society, the greater the probability that it is an open liberal society.

Proposition: The greater the degree of education and the lesser the degree of social mobility that exists in a society, the greater the probability that it is a feudal capitalist society.

Proposition: The lesser the degree of education and the lesser the degree of social mobility that exists in a society, the greater the probability that it is a society with 'closed' social structures.

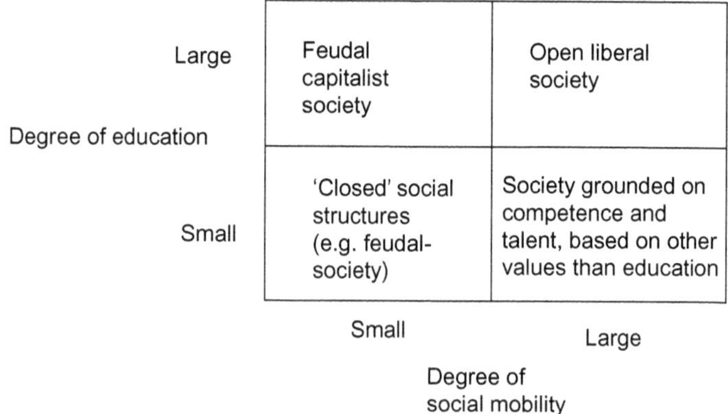

Figure 5.2 Competence transformation related to the workplace of the future: A typology.

Proposition: The smaller the degree of education and the greater the degree of social mobility that exists in a society, the greater the probability that it is a society based on values other than education which forms the ideological basis of the society.

Practical utility related to competence transformation

New technology affects the market, influencing the choices decision-makers have. Decision-making is made at several levels. Decision-makers at the individual level demand goods and services. The market for these decision-makers is largely related to real choices. When the choices are increased, the market is viewed as being freer and more inclusive for more and more people. If the market changes so that the choices are restricted, then the market will be perceived as less free. The new AI and robotics technology affects the extent of free choice in the market. It is precisely in this context that many people will be negatively impacted by the robot shock. If the further development and application of AI technology leads to increasing automation, many people will feel that their purchasing power and consumer position will have been diminished in the new innovation economy. For these people, this will feel like their choices in the market have been restricted. When choices are restricted, it is probable that the same people will feel that their freedom has also been restricted.

At the organizational level, this development will be viewed differently. Robot automation will reduce labour costs and increase productivity and profits for companies. In other words, what may be viewed as a negative development at the level of the individual, will, as a general rule, be perceived as a positive development at the organizational level.

At the social level, the tension between the individual level and the organizational level can be resolved by using an ideology that promotes the interests of people or organizations. Some ideologies will attempt to propose that if things go well at the organizational level, then this will be positive for people as a whole. However, this would be an ideology that is motivated by promoting someone's interests at the expense of others.

For the individual employee, the changes in the market will be a signal that their social mobility and the social mobility of their children is under threat. In other words, robot automation will result in a loss of jobs. The perceived threat to social mobility can be resolved by the individual employee, if he/she acquires skills that are compatible with the new technology. However, such an extensive technological education at university level will not necessarily facilitate upward social mobility, that is, moving to a higher social class. On the other hand, such an education will ensure good well-paid jobs. Those people who do not react to the change signals being sent by the market may complete an extensive university education that is not compatible with the new technology. These people, if they do not have other skills that are in demand, may risk downward social mobility, that is, moving down a social class.

One consequence of the robot shock will be a narrowing of choice. This may sound like a paradox, because technological development usually leads to an expansion of people's choices. This requires some explanation. When robots take over many of the work functions, new work functions are also created. However, the new jobs will be filled by people with skills that are in demand, while others will have to find less attractive jobs. Those people with extensive technological educations, that are compatible with the new technology, will be able to find secure well-paid jobs. For these people, their choices will be increased, and they will perceive the robot shock as something positive. The other group, who lost their jobs due to robot automation, will very likely end up in what Standing (2014, 2014a) refers to as the precariat. The precariat is not a homogeneous group, but the people in this group have one thing in common – their jobs are insecure. This group, which is increasing in numbers, will not experience that the choices in the market have increased but rather the opposite – that their freedom to choose has been restricted.

On the basis of the above description, we have formulated the following proposition: The more impact that robot shock has on the labour market, the more likely we are to experience restrictions on our individual freedoms. This proposition about limited choices is a crucial assumption related to the robot shock. For most people, the likely result of robot shock will be a descent down the social ladder into the growing ranks of the precariat. A few people will benefit from robot shock, however. This minority will enjoy increased influence and wealth with a very few of them ascending to the ranks of the elite.

Previously, a distinction could be made between those with a good university education and those without such an education. On the basis of this distinction, one could say something about the probability of future success or lack of success in the labour market. However, in the future, this may very likely change. The reason is that even if you have an extensive university education, it is not certain that this will lead to success in working life. It is this insight that emerges in Ainley's research and which is blatantly framed in the title of his book: *Betraying a Generation: How Education is Failing Young People* (Ainley, 2016). The distinction that says something about the individual's position in the future labour market is still linked to having, or not having, a good education that is in demand. However, this type of education will be one that is compatible with the new technology. In other words, those with an education related to the new technology will have the greatest probability of success in the future labour market.

The development of globalization from the 1980s until around the 2020s has strongly integrated many areas, such as the economic, political, social, cultural, educational and technological fields. This integration has also led to competition where some areas have become more important. For instance, globalization has had a direct impact on the field of education. This is evident if we consider the fact that hundreds of thousands of Chinese students have been able to attend American universities. On graduation, many of these students were able to return to China and join the middle class (Li, 2021; Rocca, 2018). In the US, competition with China led to the loss of many secure and well-paid industrial and middle-class jobs (Reeves, 2018). This consequence of globalization led to feelings of resentment, frustration, powerlessness and alienation spreading amongst large sections of the population (Frank, 2004, 2020; Carney, 2020). Gradually, this resentment, and the technological competition between the US and China, resulted in the American government restricting Chinese access to American markets. Amongst other things, the US government prevented Chinese technology from competing with American technology in various ways. This was not because the Chinese technology was better than American technology, but because the Chinese products were cheaper than the American products, due to the fact that high-tech labour costs were lower in China than in the US and Europe. Therefore, the US government used the argument about 'the risk to national security' when they prevented Chinese companies, such as Huawei, from participating in the development of 5G networks. The American technological containment policy towards China continued in August 2022 when the Chips and Science Act was passed. This made it almost impossible for Chinese technology to compete on an equal footing with the US . It can be said that this law also heralded the end of globalization and the beginning of de-globalization (Johannessen, 2023b).

De-globalization will have major consequences for future workplaces. What is highly likely, however, is that the high-tech competition between China and the US will increase in strength, because both countries realize that the country that dominates high-tech development, such as in the fields of artificial intelligence, Big Data, gene-editing and associated technologies, will become the dominant power in the future; this applies to both economic and military dominance. Consequently, there is much evidence that education that is compatible with the new technology will become more important in the future in both China and the US. The competition between China and the US will affect the development of globalization and eventually reveal whether de-globalization will increase military tensions between the two countries. Moreover, the high-tech competition will also determine who will be the winners and losers in the future labour market.

On the basis of the above narratives, description, theoretical reflections and the review of practical utility, we have developed a Boudon-Coleman diagram, which

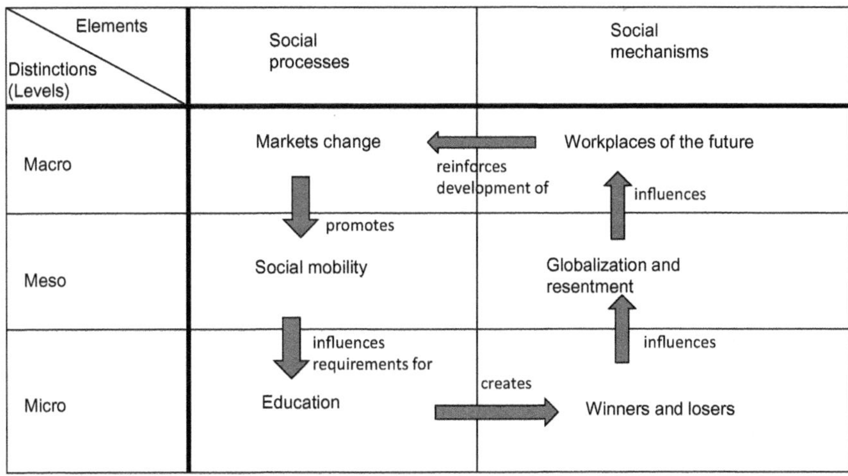

Figure 5.3 Competence transformation and the workplace of the future: A Boudon-Coleman diagram.

says something about competence transformation and the workplace of the future (Figure 5.3).

Systemic connections related to competence transformation

There are close relationships at all levels and all the functional areas in the innovation economy. These relationships are related to four elements in particular. These elements are:

1. Demands of the market
2. Globalization and its consequences
3. Increasing inequality
4. Reduced social mobility in the West but increased social mobility in China, although economic inequalities are also growing very strongly

The demands of the market in both the West and in China are related to technological development in general, and artificial intelligence in particular. The social mechanism that promotes technological development is education. Both the US and China have high quality education systems that focus on technological development. However, it may appear that the US perceives China to be a growing technological threat. This is evident if we consider the Chips and Science Act[20] which was adopted in the US in August 2022. If the US had not felt threatened by China's progress in high-tech development, especially AI technology, then it is improbable that such legislation would have been adopted. In other words, we might propose that competence transformation is the underlying dimension for the four systemic connections we have shown above.

The four systemic connections which are here related to competence transformation take into account both the internal and external influences of the economic, social, political and social developments in the innovation economy. This competence transformation will hasten the robotization of most processes in the workplace and is an essential part of the robot shock we have described throughout this book.

In this context, economic inequality and social mobility will be the greatest consequences of the robot shock for the individual. For businesses and the institutional level, globalization and market development will have the greatest consequences. Earlier in this chapter, we discussed that the world, it seems, is now heading towards de-globalization. If this assumption is correct, then it will promote increased tensions between the US and its allies on the one hand and China and its allies on the other. However, we do not believe that it is predetermined that this development will lead to increasing military tensions between the US and China. They have close economic ties, and both countries are dependent on each other for economic growth and will most probably avoid the risk of escalating tensions. In other words, both countries seek stability as a foundation for their geopolitical ambitions. On the other hand, they have a hegemonic power rivalry throughout the Asia-Pacific. In this context, one can refer to 'Thucydides Trap', which proposes that a conflict can develop when a rising great power (e.g., China) challenges the dominant great power (e.g., the US) (Thucydides, 2009). This historical supposition has been described and analysed by Allison (2018). If it turns out that the growing tensions lead towards a military confrontation between the two superpowers, then one might say that the robot shock has really led to a situation that has gotten completely out of control. In this context, it should be remembered that the world's economic dependence on microchips and AI technology is now so great that economic development can become completely reliant upon these resources, that is, chips production. These chips are largely produced in Taiwan, China, South Korea and Japan. Taiwan is in a special position, because the Taiwan Semiconductor Manufacturing Company Limited (TSMC), which is the world's most valuable semiconductor company, is located in Taiwan.[21] Taiwan makes 65 per cent of the world's semiconductors and almost 90 per cent of the advanced chips.[22] In this light, this can partly explain the rising tensions between the US and China, regarding the relations with Taiwan. In other words, the development of this situation may determine the futures of both countries with regards to economic growth, as well as with regard to military capabilities.

The increased market demand for chips and artificial intelligence technology is creating tensions not only between China and the US but also between the US's allies and China. For instance, in 2021, China replaced the US as Germany's main trading partner.[23] Germany's economic dependence on China will undoubtedly affect the relations between the three countries.

At the individual level, the largest consequences of the development described above is related to the growing economic inequality, which is evident in the US and China. Between the 1980s and 2022, the incomes of both the working and middle class in the US were negatively impacted; moreover, the middle class has gone into decline (Reeves, 2018). In China, the working class and the middle class have also been affected but in exactly the opposite way. The wages of Chinese industrial workers has increased, although their wage levels are still much lower than for American industrial workers. The middle class in China is growing sharply and around 2020 numbered approximately 700 million people (Johannessen, 2023a, 2023b). However, the economic inequality between the working class and the middle class is increasing and also between rural and urban areas (Chuliang & Sicular, 2013; Fan et al., 2009).

Globalization has led to the emergence of many areas with economic problems.[24] This development cannot be blamed on the market. For several hundred years, possibly since the 17th century in Holland, it has been shown that a free market stimulates the economy. However, when the market is corrupted by oligopolies and monopolies, the freedom of the market is sacrificed in favour of profits for the few. In this context, globalization is playing a decisive role in corrupting the market. Market freedom is restricted, competition is reduced and monopoly-like 'price-deals' are implemented in order to increase profits (Krugman, 2019). It is precisely here, at the centre of capitalism, that the free market is taken over by robots. These robots give additional power and income to those who already have considerable power and high incomes. However, it requires very large resources to develop and use these intelligent robots. The costs of this development is so great that it is only the multinational companies that are able to make such investments. By developing and adopting these robots in their production, distribution and consumption processes, these companies grow even more, increasing their productivity, while reducing labour costs. In this way, they restrict the free market and introduce a form of market regulation that serves their own interests (Baldwin, 2019; Sachs, 2020).

The applications of intelligent robots integrated with artificial intelligence have already been a game changer in many markets (Sadler et al., 2019). Slobodian (2020) says that globalization led to the development of neoliberalism. We agree with this claim but would like to add that neoliberalism also reinforced globalization, economic inequality and downward social mobility in the West. On the other hand, globalization promoted upward social mobility in China. Globalization and neoliberal policies also served the interests of the upper class in the West, while promoting China's economic growth. In a broader perspective, and in an American context, one might say that eventually 'the chickens came home to roost'. The negative consequences of neoliberalism and globalization for the US working and middle class and the positive consequences for the Chinese middle class, ultimately had its political reaction in the US. Both President Trump and President Biden pursued protectionist and nationalist economic policies aimed at China. In other words, from a US standpoint, it became clear that neoliberal ideology and globalization no longer served American interests, which led to an increasing de-globalization (Johannessen, 2023b). However, it does not seem that this de-globalization will lead to less economic inequality or increasing social mobility in the West. On the other hand, de-globalization is closely linked to the further development and application of AI and robotics technologies. This development,

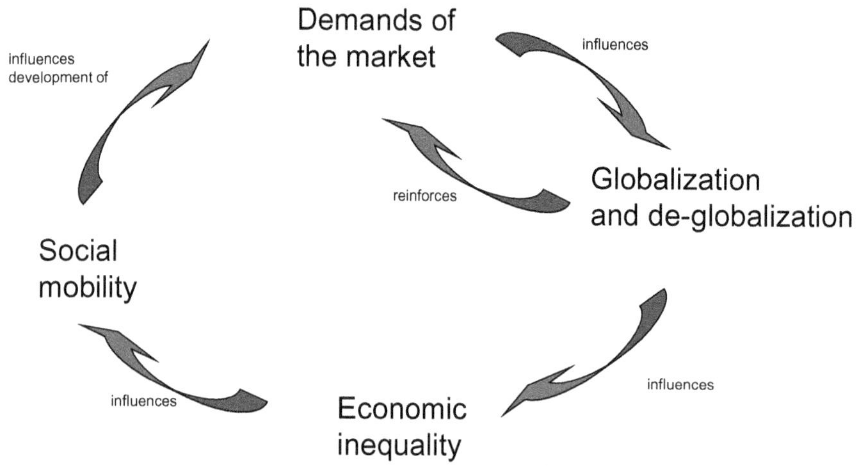

Figure 5.4 Systemic relationships related to competence transformation: A loop diagram.

as discussed above, is creating global tensions in relation to gaining control of the resources in the development of these technologies. More specifically, the production of microchips, which are crucial to the various applications of AI technology. These chips are largely produced in Taiwan, which can partly explain the rising tensions today between the US and China, regarding relations with Taiwan. In this context, one might say that the rising US–China tensions regarding Taiwan might finally close the Thucydides' Trap.

In this context, the robot shock may turn out to be a factor in the tensions that are building up in the South China Sea in general, and around Taiwan in particular. If the US and China are unable to resolve the conflicts regarding Taiwan, regardless of whether this is considered from an economic-technological point of view or a value perspective, then these tensions can trigger conflicts that no one will benefit from (Allison, 2018; Ma, 2021). If such tensions evolve into an open military conflict, then the robot shock will truly have become a shock that will have catastrophic consequences.

Based on the above narratives, description, theoretical reflections, the practical utility value and the systemic connections, we have developed a loop diagram showing the relationships discussed above (Figure 5.4).

Sub-conclusion related to transformation in competence

In this chapter, we have examined the following question: What is the potential impact of transformation in competence on future employment trends?

The short answer is that competence transformation is a necessary prerequisite for surviving the robot shock. The more in-depth answer can be understood from different perspectives and at different levels. The levels we will elaborate on in the following are: the individual level, the organizational level, the national level and the global level. The perspectives we look at are related to talent and competence.

The individual level

For the individual, setting goals and achieving them after hard work is necessary if he/she wishes to develop and realize his/her talent. For example, this applies to education and sports. Developing special talents may be one way to avoid the requirement to complete a long education compatible with the new technology in order to survive the robot shock. However, it is doubtful whether this 'talent strategy' can be used by most people as a general strategy for surviving the robot shock. Although it can be argued that everyone has some talents that they can develop, it is doubtful whether there are that many who can make a living from developing their talents. Therefore, education, which also promotes the individual's talents, is a necessary prerequisite for surviving the robot shock.

Much is written about 'the war for talent'. Therefore, the individual should reflect on how to plan a strategy to cultivate in-demand skills that are needed in the labour market. Although it is important for the individual to develop their talents, for the vast majority of people, it will be more effective to focus on competence development rather than cultivating their particular talents. In this context, this is where education is important. On the other hand, not everyone has special talents or equal opportunity to complete a good education. In other words, educational equality and talents do not constitute the core of competence transformation; competence transformation is mainly concerned with developing a strategy to survive the robot shock. The point of mentioning talent contra education here, which of course need not be opposites, is that these two social mechanisms overshadow all other social mechanisms for most people in order to achieve social mobility. There are also other social mechanisms that can enable the individual to survive the robot shock. However, these social mechanisms only exist for a minority in the upper class. These people have access to the necessary resources, such as wealth, special social relationships, support apparatus, etc.

Not everyone has the same opportunities or prerequisites to develop their skills or talents. Our point is that even if this is correct, developing one's competence will be a safer strategy than developing one's talents. For some, it may of course be possible to do both.

The organizational level

For organizations, it will also be important to develop talents within the organization. At the same time, it will be crucial for an organization's progress to hire and develop those people with the skills that are in demand. Employees will also need to update their skills. However, this will not apply to all employees. In other words, in relation to the robot shock we are investigating here, the people who do not have competence that is compatible with the new technology will be amongst the first to lose their jobs. Organizations are of course most focused on performance, profit, market share, etc.; consequently, they will not prioritize those employees who are unable to meet the demands of the new technology. Therefore, the challenges regarding employees without the necessary skills will most probably be addressed by the government authorities.

The national level

To compete in the innovation economy and promote economic growth, countries will need to prioritize education and research in the new technology. Without such an educational policy, it may take too long to acquire the necessary national competence to be able to compete with other nations in an age where the development and application of AI and robotics technologies are becoming crucial for economic growth. No matter how well prepared a country is for the upcoming technological competition, there will always be some people and businesses that, for various reasons, are unable to meet the demands for new skills. In this context, governments will need to create efficient safety nets to support those people and businesses that are unable to make the transition to new skills.

Loss of opportunities at all levels at a time when it becomes important how quickly businesses and countries are able to secure a competitive position can be decisive for further development. Therefore, when competence transformation is crucial, we will most likely see a development in most businesses and countries, where there will be an increased demand for competences, talents, loyalty, relational strength and so on, etc. The increasing competition to survive when the robot shock makes itself felt will apply not only to individuals, organizations and nations but also at the global level.

The global level

At the global level, especially regarding the relationship between the US and China, the country that takes the lead in the development of the new technologies, such as artificial intelligence, will most probably take the global leadership in future decades. Decisive in this context will be the ability to secure the necessary resources to develop and apply these technologies.

If we consider the aforementioned four levels and the perspectives we have emphasized at the various levels, we can develop the following proposition as an answer to the question we have examined in this chapter: As demand grows for higher levels of competence transformation, more stringent performance criteria will most likely be applied across the system as a result.

Notes

1 https://en.wikipedia.org/wiki/First-mover_advantage
2 Acemoglu et al., 2015; Atkinson, 2015; Dorling, 2015, 2019; Eubanks, 2017; Milanovic, 2016; Picket, 2017; Piketty, 2014, 2016.
3 https://da.wikipedia.org/wiki/Adam_Smith
4 https://da.wikipedia.org/wiki/Paul_Samuelson
5 Science, Technology, Engineering, Mathematics. Biological competence is largely related to S (science) in STEM.
6 https://en.wikipedia.org/wiki/Open_Letter_on_Artificial_Intelligence
7 https://www.brookings.edu/wp-content/uploads/2021/11/111621-BROOKINGS-AUTOMATION-TRANSCRIPT.pdf

8 Robert Solow's original statement was: "You can see the computer age everywhere but in the productivity statistics" (ref i: https://en.wikipedia.org/wiki/Productivity_paradox).
9 Brynjolfsson & McAfee, 2011, 2014; Brynjolfsson & Saunders, 2013.
10 https://en.wikipedia.org/wiki/Productivity_paradox
11 https://en.wikipedia.org/wiki/CHIPS_and_Science_Act
12 https://en.wikipedia.org/wiki/Meritocracy
13 https://en.wikipedia.org/wiki/General-purpose_technology
14 https://en.wikipedia.org/wiki/CHIPS_and_Science_Act
15 Chuliang & Sicular, 2013; Fan et al., 2009; Kanbur & Zhang, 2009; Knight et al., 2013; Liu, 2016; Shi et al., 2013;
16 Antonelli, 2009; Bessen, 2020; Frey, 2019, 2020; Goldin & Katz, 2008; Elkington, 2020; Ellul, 1967, 2018; Habakkuk, 1962; Kumar et al., 2021; Lipsey & Karlaw, 2005; Sachs, 2020; Susskind, D., 2020.
17 https://en.wikipedia.org/wiki/CHIPS_and_Science_Act
18 https://en.wikipedia.org/wiki/Artificial_intelligence_industry_in_China
19 https://en.wikipedia.org/wiki/Regional_Comprehensive_Economic_Partnership
20 https://en.wikipedia.org/wiki/CHIPS_and_Science_Act
21 https://en.wikipedia.org/wiki/TSMC
22 https://www.voanews.com/a/race-for-semiconductors-influences-taiwan-conflict-/6696432.html
23 https://www.destatis.de/EN/Themes/Economy/Foreign-Trade/trading-partners.html
24 Acemoglu & Autor, 2011; Dorling, 2019; Fan et al., 2009; Liu, 2016; Picket, 2017; Piketty, 2014; Shi et al., 2013; Shipler, 2005; Standing, 2016; Stilwell, 2019.

References

Acemoglu, D. & Autor, D.H. (2011). Skills, tasks and technologies: Implications for employment and earnings, in Card, D. & Ashenfelter, O. (eds.), *Handbook of labor economics*, vol. 4, Elsevier, Amsterdam, pp. 1043–1171.

Acemoglu, D.; Suresh, N.; Pascual, R. & Robinson, J.A. (2015). Democratic redistribution, and inequality (Chapter 21), in Atkinson, A.B. & Bourguignon, F. (eds.), *Handbook of income distribution*, vol. 2, Elsevier, London, pp. 1885–1966

Ainley, P. (2014). Follow your dreams and attend to universities if possible, *Latitude*, 21 December.

Ainley, P. (2016). *Betraying a generation: How education is failing Young people*, Policy Press, Bristol.

Allison, G. (2018). *Destined for war: Can America and China escape Thucydides trap?* Scribe, London.

Antonelli, C. (2009). The economics of innovation: from the classical legacies to the economics of complexity, *Economics of Innovation and New Technology*, 18 (7): 611–646. DOI: 10.1080/10438590802564543. To link to this article: http://dx.doi.org/10.1080/10438590802564543.

Atkinson, A.B. (2015). *Inequality: What can be done*, Harvard University Press, Cambridge.

Baldwin, R. (2019). *The globotics upheaval: Globalization, robotics and the future of work*, Oxford University Press, Oxford.

Batzhan, A.; Alpysbayeva, S. & Kapsalyamova, Z. (2020). The belt and road initiative: Case of Kazakhstan, in Wah, C.Y.; Menkhoff, T. & Low, L. (eds.), *China's belt and road initiative: Understanding the dynamics of a global transformation*, World Scientific, London, pp. 71–97.

Beer, S. (1994). *Beyond dispute: The invention of team syntegrity*, Wiley, New York.

Beer, S. (1995). *Diagnosing the system for organizations*, John Wiley & Sons, London.

Beeson, M. & Li, F. (2016). China's place in regional and global governance: A new world comes into view, *Global Policy*, 7 (4): 491–499.
Berlie, J.A. & Hung, S. (2020). The greater Bay area and the role of Hong Kong and Macau SARs in the belt and road initiative, in Berlie, J.A. (ed.), *China's globalization and the belt and road initiative*, Palgrave, London, pp. 77–100.
Bessen, J. (2020). Attitudes to technology: Part 1, in Skidelsky, R. & Craig, N. (eds.), *Work in the future*, Palgrave, London, pp. 83–88.
Brown, K. (2018). *Chinas dream*, Polity, London.
Bruder, J. (2017). *Nomadland: Surviving America in the twenty-first century*, W.W.Norton, New York.
Brynjolfsson, E. & McAfee, A. (2011). *Race against the machine*, Digital Frontier Press, New York.
Brynjolfsson, E. & McAfee, A. (2014). *The second machine age*, W.W. Norton & Company, New York.
Brynjolfsson, E. & Saunders, A. (2013). *Wired for innovation: How information technology is reshaping the economy*, The MIT Press, London.
Bublitz, H. (2022). *Die verborgenen Codes der Erben Über die soziale Magie und das Spiel der Eliten*. Transcript, Berlin.
Carney, T.P. (2020). *Alienated America*, Harper, New York.
Chetty, R.D; Grusky, M.; Hell, N.; Hendren, R. Manduca, R. & Narang, J. (2017). The fading American dream: Trends in absolute income mobility since 1940, *Science*, 356 (6336): 398–406.
Chomsky, N. (2016). *Profit over people: War against people*, Piper, Berlin.
Chomsky, N. (2016a). *Who rules the world*, Hamish Hamilton, London.
Chuliang, L. & Sicular, T. (2013). Inequality and poverty in rural China, in Sji, L.; Sato, H. & Sicular, T. (eds.), *Rising inequality in China: Challenges to a harmonious society*, Cambridge University Press, Cambridge, pp. 197–229.
Clinton, H. (2011). America's Pacific century, *Foreign Policy*, 189, 11 October. https://foreignpolicy.com/2011/10/11/americas-pacific-century/#cookie_message_anchor
Deng, X. (1994). *Selected works of Deng Xiaoping, vol III (1982–1992)*, Foreign Language Press, Beijing.
Dorling, D. (2015). *Injustice: Why social inequality still persists*, Policy Press, London.
Dorling, D. (2019). *Inequality and the 1%*, Verso, London.
Doshi, R. (2021). *The long game: China's grand strategy to displace American order*, Oxford University Press, Oxford.
Drache, D.; Kingsmith, A.T. & Qi, D. (2019). *One road many dreams: China's bold plan to remake the global economy*, Bloomsbury, New York.
Elkington, J. (2020). *Green Swans: The coming boom in regenerative capitalism*, Fast Co PR, New York.
Ellul, J. (1967). *The technological society*, Vintage, London.
Ellul, J. (2018). *The technological system*, Wipf and Stock, London.
Eubanks, V. (2017). *Automating inequality*, St. Martin's Press, New York.
Fan, S.; Kanbur, R. & Zhang, X. (2009). Regional inequality in China: An overview, in Fan, S.; Kanbur, R. & Zhang, X. (eds.), *Regional inequality in China*, Routledge, London, pp. 1–13.
Ford, M. (2016). *The rise of the robotics, technology and the threat of mass unemployment*, Oneworld, New York.
Ford, M. (2021). *Rule of the robots*, Basic Book, London.
Frank, T. (2004). *What's the matter with Kansas? How conservatives won the Heart of America*, Metropolitan Press, New York.
Frank, T. (2020). *People without power*, Scribe, London.
Frey, C.B. (2019). *The technology trap: Capital, labor, and power in the age of automation*, Princeton University Press, Princeton.

Frey, C.B. (2020). Attitudes towards technology: Part II, in Skidelsky, R. & Craig, N. (eds.), *Work in the future: The automation revolution*, Palgrave, London, pp. 89–97.
Gerstle, G. (2022). *The rise and fall of neoliberal order*, Oxford University Press, Oxford.
Goldin, C. & Katz, L.F. (2008). *The race between education and technology*, The Belknap Press, New York.
Habakkuk, J. (1962). *American and British technology in the nineteenth century, the search for labor savings inventions*, Cambridge University Press, Cambridge.
Harari, Y.N. (2022). Introduction, in Huxley, A. (ed.), *Brave new world*, Vintage Classic, London, pp. vii–xi.
Iversen, T. & Soskice, D. (2019). *Democracy and prosperity*, Princeton University Press, Princeton.
Jain, R. (2021). *China's soft power and higher education in South Asia*, Routledge, London.
Johannessen, J-A. (2020). *The workplace of the future*, Routledge, London.
Johannessen, J-A. (2020a). *Automation, innovation and economic crises: Survival the fourth industrial revolution*, Routledge, London.
Johannessen, J-A. (2020b). *Artificial intelligence, automation and the future of competence at work*, Routledge, London.
Johannessen, J-A. (2021). *China's innovation economy: Artificial Intelligence and the new silk road*, Routledge, London.
Johannessen, J-A. (2021a). *Artificial intelligence, automation and ethics in the innovation economy*, Routledge, London.
Johannessen, J-A. (2023). *Feudal capitalism in the innovation economy*, Routledge, London.
Johannessen, J-A. (2023a). *The new silk road and the innovation economy in China*, Routledge, London.
Johannessen, J-A. (2023b). *De-Globalization in the innovation economy:* ***Technological innovations trigger conflict between China and the United States***, Routledge, London.
Kahneman, D. (2011). *Thinking fast and slow*. New York: Allen Lane.
Kahneman, D., & Tversky, A. (1979). An analysis of decision under risk. *Econometrica, Journal of the Econometric Society,* 47 (2): 263–292.
Kahneman, D., & Tversky, A. (2000). Prospect theory: An analysis of decision under risk, in Kahneman, D. & Tversky, A. (eds.), *Choices, values and frames*, Cambridge University Press, Cambridge, pp. 17–43.
Kanbur, R. & Zhang, X. (2009). Which regional inequality? In Fan, S.; Kanbur, R. & Zhang, X. (eds.), *Regional inequality in China*, Routledge, London, pp. 15–31.
Knight, J.; Sicular, T. & Ximing, Y. (2013). Educational inequality in China: The intergenerational dimension, in Shi, L.; Sato, H. & Sicular, T. (eds.), *Rising inequality in China*, Cambridge University Press, Cambridge, pp. 142–197.
Kotkin, J. (2020). *The coming of neo-feudalism: A warning to the global middle class*, Encounter Books, New York.
Krugman, P. (2019). Globalization: What did we miss? In Catao, L.A.V. & Obstfeld, M. (eds.), *Meeting globalization's challenges*, Princeton University Press, Princeton, pp. 113–121.
Kumar S., Aggarwal, V. & Gupta, S. (2021). Artificial Intelligence and Nanotechnology: A super convergence, in Bhargava, C. & Kumar Sharma, P. (eds.), *Artificial intelligence: Fundamentals and applications*, Routledge, London, pp. 1–9.
Li, C. (2021). *Middle class Shanghai: Reshaping US-China engagement*, Brooking Institution Press, Washington, DC.
Li, R. (2020). *Artificial intelligence revolution: How AI will change our society, economy and culture*, Skyhorse Publishing, New York.
Lipsey, R. G. & Carlaw, K. I. (2005). *Economic Transformations: General purpose technology and long term economic growth*, Oxford University Press, Oxford.
Liu, Y. (2016). *Higher education, meritocracy and inequality in China*, Springer, London.
Ma, W. (2021). *The digital war*, Wiley, New York.

Milanovic, B. (2016). *Global inequality*, The Belknap Press, New York.
Petras, J. & Veltmeyr, H. (2011). *Beyond neoliberalism: A word to win*, Routledge, London.
Picket, K. (2017). Foreword, in Brown, R. (ed.), *The inequality crises*, Policy Press, London, pp. vii–viii.
Piketty, T. (2014). *Capital in the twenty-first century*, The Belknap Press of Harvard University Press, Boston.
Piketty, T. (2016). *Chronicles: On our troubled times*, Viking, London.
Przeworski, A. (2019). *Crises of democracy*, Cambridge University Press, Cambridge.
Reeves, R.V. (2018). *Dream Hoarders: How the American upper middle class is leaving everyone else in the dust. Why that is a problem and what to do about it*, Brooking Institution Press, Washington, DC.
Rocca, J-L. (2018). *The making of the Chinese middle class: Small comfort and great expectations*, Palgrave, London.
Sachs, J.D. (2020). *The ages of globalization: Geography, technology and institutions*, Colombia University Press, New York.
Sadler, M.; Regan, N.; Hassabis, D. & Kasparov, G. (2019). *Game changer: AlphaZeros ground-breaking chess strategies and the promise of AI*, New in Chess, London.
Sandel, M.J. (2020). *The tyranny of merit: What's become of the common good?* Allen Lane, New York.
Schwab, K. (2016). *The fourth industrial revolution*, World Economic Forum, Geneva.
Schwab, K. (2018). *Shaping the fourth industrial revolution*, World Economic Forum, Geneva.
Sennett, R. (1998). *The corrosion of character: Personal consequences of work in the new capitalism*, W.W. Norton & Company, New York.
Sennet, R. (2004). *Respect*, Norton, New York.
Sennet, R. (2006). *The culture of the new capitalism*, Yale University Press, London.
Sennett, R. (2008). *The craftsman*, Yale University Press, New Haven, CT.
Sennett, R. (2013). *The rituals, pleasures and politics of cooperation*, Penguin, London.
Shi, L.; Sato, H. & Sicular, T. (2013). Rising inequality in China: Key issues and findings, in Shi, L.; Sato, H. & Sicular, T. *Rising inequality in China*, Cambridge University Press, Cambridge, pp. 1–44.
Shipler, D. (2005). *The working poor*, Vintage, New York.
Standing, G. (2014). *The precariat: The new dangerous class*, Bloomsbury Academic, New York.
Standing, G. (2014a). *A precariat charter*, Bloomsbury, London.
Standing, G. (2016). *The corruption of capitalism: Why rentiers thrive and work does not pay*, Biteback Publishing, London.
Standing, G. (2019). *Plunder of the commons, A manifesto for sharing public wealth*, Pelican, New York.
Stiglitz, J.E. (2015). *Rewriting the rules of the American Economy: An agenda for shared growth and prosperity*, W.W. Norton and Co., New York.
Stiglitz, J.E. (2019). *People, power and profits*, Penguin, London.
Stilwell, F. (2019). *The political economy of inequality*, Polity, London.
Susskind, D. (2020). *A world without work: Technology, automation and how we should respond*, Allen Lane, London.
Thucydides. (2009). *The Peloponnesian war*, Oxford University Press, Oxford.
Vogel, E.F. (2011). *Deng Xiaoping and the transformation of China*, Harvard University Press, Cambridge, MA.
Weeks, J.F. (2014). *Economics of the 1%: How mainstream economics serves the rich, obscure reality and distorts policy*, Anthem Press, New York.
Weick, K.E. (1979). *The social psychology of organizing*, Wiley, New York.
Zuboff, S. (2019). *The age of surveillance capitalism*, Profile Books, London.

6 Waves of black swan events

Core idea in this chapter

Polycrises increase the complexity of social systems.

Key points in this chapter

- Innovations emerge where there is relative falling productivity, increasing relative costs, limited access to expertise, falling relative quality, a high rate of the spread of knowledge and where knowledge has the potential to be transformed into new technology.
- We don't know what jobs will be like in the future, but we do know about the social mechanisms that will drive the creation of these jobs.
- We have developed nine strategies for tackling Maginot mentalities.
- When it comes to systemic innovations, it may be useful to draw a map that is capable of changing the terrain it represents.
- It is not adaptability but readiness to change that is relevant for tackling systemic innovations.
- The distinction between evolutionary processes in nature and in social systems is found in the relationship between random selection and deliberate selection.
- Contextual competence leads to systemic innovations. There are many examples of systemic innovations that have been created by contextual competence: Biotechnology, nanotechnology, gene-editing, molecular computing, AI protein folding, brain-computer interfaces (BCIs) and artificial intelligence (AI), AI and game development, AI and data analysis, AI body sensors, AI environmental sensors, AI and medicine and so on.

Introduction

The problem we investigate in this chapter concerns the most probable social consequences of artificial intelligence. Specifically, we examine the following question: What is the potential impact on future employment trends of the fact that the pace of change may outstrip our ability to adapt? A key objective of this chapter is to develop strategies to survive 'robot shock' and in particular the global

DOI: 10.4324/9781003567325-6

polycrisis that appears to be affecting the innovation economy as we enter the Fourth Industrial Revolution.

When the world changes slowly, we can manage to adapt and thus change our behaviour. However, if situations arise where the environment changes very quickly, completely different types of situations arise. In such situations, we can say metaphorically that flocks of black swan events occur.[1] Another way of understanding this is to use the term polycrisis. A polycrisis is the simultaneous occurrence of several catastrophic events, such as the economic crisis, the climate change crisis, the environmental crisis, the refugee crisis, the inequality crisis, the energy crisis, the food crisis, the health crisis, etc. Such polycrises are rare in history. On the other hand, it may be correct to say that such crises occur in the wake of industrial revolutions. The crises that emerge in connection with industrial revolutions are mainly related to economic, social and technological changes that occur more or less simultaneously, viewed from a historical perspective. If in addition to such crises, other crises also occur, as mentioned above, such as the climate change, inequality, environmental, energy and food crises, then such a situation can be described as a global polycrisis.[2] Moreover, We say that these crises are systemically linked, because changes affecting one of these crises will have an impact on the other simultaneous crises. A global polycrisis will exacerbate geopolitical risks.[3]

It is not what we know about these interconnected crises that is frightening; this is already worrying enough, but what we do not know about such a polycrisis, especially a global polycrisis, is extremely menacing. Perhaps the most terrifying thing is the realization that we don't even know what we don't know. In such a situation, we don't even know where to start looking for knowledge to solve problems, because we don't know which questions to ask. In such a situation, we need to learn, as individuals, organizations, and society, how to discover relevant and sufficient knowledge to deal with these interconnected crises. What we do know is that the crises are interconnected and systemic. This insight then gives us a tool to begin devising strategies to deal with these interconnected crises. This tool is systemic thinking, because this way of understanding such crises is also compatible with how these crises appear and emerge.

What is surprising is not that we cannot predict what the workplaces of the future will look like, but that we are unable to realize that the interconnected crises will create an emergent which then will change all the assumptions about how our workplaces are organized. Although we cannot predict what the workplaces of the future will look like, this does not mean that we do not know what will affect them most, and thus change the way we organise, produce, distribute and consume. We know that artificial intelligence and intelligent robots will affect most industries. This will apply to journalism, the construction industry, the medical industry, the pharmaceutical industry, the cultural sector, the military and so on. Of course, we do not know exactly how these industries will be affected. In contrast, we do know the following:

1. When AI-powered automation increases, costs will also increase initially.
2. The relationship between costs and automation can be understood as an inverted U. This means that as automation increases, costs will rise to the highest point on this inverted U. Then costs will fall.
3. This also implies that labour productivity will initially decline when automation increases, because costs increase.
4. When we reach the top point of the inverted U, productivity will begin to rise, slowly at first and then exponentially.
5. The explanation for the relationship mentioned in points 1–4 above is coupled to learning and a more usable technology that changes through trial and error in practice.
6. We also know that artificial intelligence and intelligent robots influence:
    a. the employee's work situation,
    b. an organization's production, distribution and consumption,
    c. a nation's strategic visions in several areas, e.g., competence development, and
    d. the geopolitical situation, amongst others, relations between China and the US.

The changes we have indicated above are related to a time perspective. This time perspective will vary for different industries. With regard to the development and application of artificial intelligence, areas that are perceived as being related to national security, such as the military, will probably receive more focus than other industries. Historical knowledge indicates that the rate of spread of new technology, from the industry that is at the forefront of development to other industries, will be great in the industries that have:

a. falling relative productivity,
b. increasing relative costs,
c. low access to expertise,
d. declining relative quality,
e. high rate of spread of knowledge, and
f. where knowledge has the potential to become new technology.

This insight is based on the various works of P. F. Drucker (1999, 1999a, 2005, 2007). The above factors imply that the health sector, where there is a growing demand for health personnel, will increasingly use artificial intelligence (AI) in the future, more so than in other sectors. Although the health sector may not be the first sector to develop and use artificial intelligence, due to the five factors (a-f) mentioned above, this sector will be far ahead of other sectors in utilizing AI. It may be the case that AI will initially be developed and used in the military.

An important point about the relationship between automation and cost development mentioned above is that those businesses and sectors that are the first to apply the new technology will also be those that have a high learning curve in relation to using the technology. This indicates that these first-movers will gain an advantage because they can quickly increase the learning curve at the same time as they learn to adapt the new technology to the relevant work operations used within the industry and the specific workplace.

Of course, we cannot predict what future workplaces will look like. We cannot even predict what the price of petrol will be in three to four months' time. However, this does not mean that we cannot say anything about what happens to employees, organizations and society when the pace of change increases faster than we can adapt to it. When such a situation arises, we know that most people will resist changes (Kahneman & Tversky, 1979, 2000). We also know from the Prospect Theory of Kahneman & Tversky what has to be done to motivate employees to participate actively in necessary organizational changes (Adriaenssen & Johannessen, 2016a). When we have this knowledge, we also know how people are likely to react when new AI technology impacts our workplaces at an ever-increasing pace.

Our inability to predict does not mean that we cannot know anything about what will happen when new AI technology enters the workplace. As mentioned above, we know where innovations will make the greatest impact in relation to points a-e above, which we have generated from P. F. Drucker's research. The point of this brief reflection is to show that even if we cannot predict in detail what the future will look like, we can say with relatively high certainty which drivers will affect this future and how people and the social system will react when the changes come quickly and in some cases faster than we can manage to adapt to before a new change appears.

Ignorance of details about how the future will take shape does not mean that we are ignorant of what drives the changes and which then will shape the workplaces of the future. What do we know?

1. We know that new technology will affect our workplaces.
2. We know that many people will oppose these changes.
3. We know how to deal with resistance to changes (Prospect Theory).
4. We know that robotic automation will first lead to higher costs and then falling costs and increased productivity.
5. We know that the top-earners will be those people who transform their competence so that it is compatible with the new technology.
6. We also know that optimism motivates and engages and that optimism can be learnt.

Black swan events are by definition impossible to predict. In contrast, even with only minimal knowledge of history and innovation, most people can predict that these kinds of black swan events will occur, even though one cannot know where or when. Black swan events could occur in at least four areas: Economics/technology, politics, culture and relational biology.

We know with a high degree of certainty that the black swans will appear as innovations and very likely as interconnected innovations or what we call systemic innovations in this chapter. They can also appear as linked crises (polycrises). We also know that these systemic innovations and interconnected crises will with a large degree of certainty increase the complexity of social systems. We also know that to cope with this complexity, creating something together, a form of co-creation, can be a useful strategy. We also know that this co-creation will require a specific form of expertise. This is a competence that is both depth-oriented and breadth-oriented. We refer to this competence here as contextual competence to distinguish it from specialist competence or expert competence. We also know that the new technology being developed is based on artificial intelligence. We know that this technology is implemented and used in what we call intelligent robots.

Although we don't know what workplaces will be like in the future, when waves of black swan events appear, we know a lot about many of the factors that drive workplace development.

Although no one could predict that COVID-19 would appear in 2020, governments knew a future pandemic was a threat, and the authorities in some countries were to some extent prepared, while in other countries they weren't well prepared.

When black swans occur, i.e., unpredictable crises, we are most concerned with what we need to do in practice to help people who are affected by the crises. There is of course nothing wrong with focusing on practical aspects and what you have to do in concrete terms. On the other hand, it might be advantageous if we theorized practice and the knowledge we have about practice. The excessive focus on practice is closely related to the fact that we emphasize what we know and make our choices of action based on this position. In other words, it makes sense to most people to start with what you know and then build your strategies on the facts. However, how did we get to the point that we had knowledge of 'the facts'? We have this practical knowledge precisely because we have investigated what we did not know about an event, action, problem, phenomenon or historical situation. It was precisely by investigating what we did not know that we could arrive at what we now know. We then use this knowledge to find practical solutions to problems and phenomena. Against this background, it can be suggested that practice is best served when we reflect upon and theorize the practical knowledge we have in order to determine if there are other approaches and strategies we might adopt that will be of more help than just the practical knowledge we already have. In other words, it can sometimes be more appropriate when we are using practical solutions to move away from this approach and theorize and conceptualize the situation we are trying to remedy and do something about. There are several historical examples where such reflection and theorizing would have been more appropriate than only taking a practical approach. When the French built the Maginot Line, a line of defensive fortifications along the French/German border, before the Second World War, they did so on the basis of the practical knowledge they had, that is, the trench warfare of World War I. During World War I, Germany had attacked France along the route along which the Maginot Line was built. In other words, the French calculated that it would be practically impossible for Germany to break down the Maginot Line with the military technology that existed at the time the Maginot Line was built. However, at the beginning of the Second World War,

Germany adopted a completely different approach to warfare than the French. They did not utilize the trench warfare practices of World War I but attacked France using *blitzkrieg* – a tactic that completely defeated the Maginot Line's purpose. In this way, they reflected on previous practice, abstracted this knowledge and came to the conclusion that the best way to win was to start from what was unknown, namely to bypass the Maginot Line.

The point of using the Maginot Line as an example that past practices and established knowledge based on facts will not always give the best results is to highlight the idea that we need to investigate what we do not know in order to improve our practices. It is of course not the historical Maginot Line that is of interest in this presentation but the fact that so-called 'Maginot Lines' can be found in many areas, such as in psychological, social, economic, technological and cultural areas. It is when we apply knowledge we already know in order to deal with practical problems and situations that so-called 'Maginot Line' practices should be theorized in order to investigate whether this knowledge is outdated or not appropriate to use in the situation at hand. The strategy for dealing with various types of 'Maginot Lines' is to examine what we know; what we don't know; what we know we don't know; and finally what we don't know, we don't know. It is in relation to these four categories of knowledge that investigations related to the many 'Maginot Lines' can give us the practical tools we need in a situation.

Practitioners may object to our recommendations here on the basis that when we are faced with a specific situation, our only option is to apply the knowledge we have available. There are few people, if any, who would disagree. The point we are trying to make is that before an event occurs, we must continuously apply the knowledge developed around the four knowledge areas in order to generate alternative knowledge that can help prevent us from running into or building our own Maginot Lines. There are many techniques for tackling one's own and others' Maginot mentalities. Some of these techniques are mentioned in the following.

1. When trying to change someone else's behaviour, the most appropriate technique may be to change one's own reactions to the other person's behaviour.
2. When dealing with the other person's psychological 'Maginot Line', it may be appropriate to think about what might have been, that is, about alternatives to our own pasts. In this way, one can address so-called tunnel vision and stimulate thinking outside the box. This technique is called counterfactual thinking.
3. To avoid 'Maginot Line thinking', it may be appropriate to reflect on collective blindness. Collective blindness may be said to be a form of collective arrogance, which results in irrational actions. Minor events slip under the radar, causing the system to not be fully aware of what is happening.
4. Doing something different or the exact opposite of what is expected can in many situations deconstruct one's own and others' psychological 'Maginot Lines'. This technique is called paradoxical intervention.
5. Examining one's own and others' values can be an approach to dismantling 'Maginot Lines'. This idea is based on North's Action Theory, which states

that: *People act on the basis of a system of rewards as expressed in the norms, values, rules and attitudes in the culture (the institutional framework).*[4]

6. To investigate which type of social response oneself and other(s) react most positively to.[5]
7. We need to be aware how the concept of 'punctuation' affects people's thinking. By punctuation[6] a distinction is drawn between cause and effect; this is done with a clear motive in mind. A causality is thus created which does not actually exist in the real world, and one is then free to discuss the effects of this cause which has been created through a process of punctuation. A sequence of a process is selected and then bracketed. In this way, we de-limit what is punctuated from the rest of the process. Figuratively, we may imagine this as a circle that is divided into small pieces; one piece of the circle is then selected and folded out into a straight line. This results in the creation of an artificial beginning and end. This beginning and end of course cannot exist in a circle but only through the process of punctuation.
8. Examine the system's social cohesion. Social cohesion refers to the strength of relationships amongst members of a social system. Social cohesion in a social system is the glue that binds the system together. Social cohesion consists of the social mechanisms that make the system durable. According to systemic thinking, relationships and actions are the glue that binds the social system together. The rationale is that relationships and the system of relationships control human behaviour. Social systems are held together (in systemic thinking) by dynamic social relations (e.g., feelings, perceptions and norms) and social actions (e.g., cooperation, solidarity, conflict and communication).
9. Examine the social mechanisms that govern the social system that you or others are a part of. Robert Merton[7] introduced the notion of social mechanisms into the field of sociology, although we can find rudiments of this in both Weber – with the Protestant ethic as an explanation for the emergence of capitalism in Europe – and in Durkheim, who uses society as an explanation for a rising suicide rate. According to Merton, social mechanisms are the building blocks of '*middle range theories*'. He defines social mechanisms as '*social processes having designated consequences for designated parts of the social structure*'.[8] In the 1980s and 1990s, Jon Elster developed a new notion of the role of social mechanisms in sociology.[9] Hedstrom and Swedberg write that, '*the advancement of social theory calls for an analytical approach that systematically seeks to explicate the social mechanisms that generate and explain observed associations between events*'.[10] It is one thing to point out connections between phenomena. It is something quite different to point out satisfactory explanations for these relationships, which is what social mechanisms accomplish. A social mechanism tells us what will happen, how it will happen and why it will happen.[11] Social mechanisms are primarily analytical constructs which cannot necessarily be observed; in other words, they are epistemological, not ontological. However, social mechanisms are observable in their consequences. An intention can be a social mechanism of action. We cannot observe an intention, but we can interpret it in light of the consequences

manifested through an action. Preferences can also function as a social mechanism for economic behaviour. We cannot observe a person's preferences, but we can interpret them in the light of the behavioural consequences that manifest themselves. Social mechanisms are, understood in this way, analytical constructs, indicating connections between events.[12] Bunge says: '... *a social mechanism is a process in a concrete system, such that it is capable of being about or preventing some change in the system as a whole or in some of its subsystems'*.[13] By 'social mechanism', here we mean those activities that promote/inhibit social processes in relation to a specific problem/phenomenon. Material resources and technology are social mechanisms of the economic subsystem; power is a social mechanism of the political subsystem; fundamental norms and values are a social mechanism of the cultural subsystem; and human relationships are a social mechanism of the social subsystem. These system-specific social mechanisms interact with each other to achieve certain goals, maintain these systems or to avoid certain undesirable conditions in the system or the outside world. The difficulty of discovering social mechanisms and distinguishing them from processes may be partly explained by the fact that social mechanisms are also processes.[14]

Narratives related to black swan events

Case letter 1: Robotic Process Automation (RPA)

In any industry or sector, the following can be performed to increase productivity:[15]

1. In step one, you ask yourself what the system is designed to do. The answer to this question is the crucial processes that define the system. In the university sector, these processes are: Teaching, research and dissemination to the field of practice.
2. In step two, each one of the processes is divided into activities.
3. In step three, one examines how robotization can be used in the various activities.
4. In step four, one examines whether there are connections between the various activities. Where there are connections, one can investigate whether it is possible to connect some of the input factors. For example, in the university sector, it is often possible to relate research to teaching. For instance, this could apply to enlisting the aid of students in the collection of data in the research process. In this way, the students learn in practice how the research process functions, while the researcher/teacher can save a lot of time and resources on collecting data.
5. In step five, one investigates whether some staff that carry out specific tasks can replace or completely take over the activities of others, so that manpower, which is a scarce resource, can be freed up. For instance, biologists who take over some of the work tasks that have been carried out by medical specialists.

6. In step six, one examines which activities can be automated using artificial intelligence, intelligent robots, intelligent algorithms, etc.
7. In step seven, you define how productivity is to be measured within the various processes defined in step one as core processes in the system, as well as the related activities. For instance, this could apply to the competence that is substituted with other competence in step five.
8. In step eight, effective measurement systems are developed to examine productivity over a period of time.

These eight steps in an RPA process can be understood as part of the system's tactical operational automation process to promote productivity. This tactical manoeuvre can be an answer to how social systems can adapt when the pace of change outstrips our ability to adapt. The point of such a procedure is to prepare the system for changes when waves of black swan events appear. These black swan events can in practice be systemic innovations that change the prerequisites for many of the work tasks carried out in various organizations. When the eight process steps have been reviewed, the employees and teams that have taken part in the survey will have gained a better overview of how to match the existing work activities with the work activities that may become a reality in the emerging future workplace.

We know that systemic innovations increase the complexity of social systems. The process we have developed above, by utilizing Stafford Beer's ideas, can be a help in making this complexity easier to deal with, while at the same time increasing the productivity of the system. Productivity can in principle be increased in two ways by using RPA. First, there can be a large increase in productivity by substituting personnel as mentioned in step five. The purpose of step five is to ease the pressure on the scarce resources and use a type of competence that is not as scarce. For example, medical specialists or psychologists can spend more time performing the tasks they are trained to do, if their routine work tasks are transferred to other professional groups who can perform these activities just as well, but at a lower cost, because their wages are lower than the specialists' wages. Moreover, productivity can be increased by the robotization and automation of some activities.

Case letter 2: Future perfect thinking

When changes happen faster than we are able to adapt to them, it is important that we at all levels, the individual, organizational and social levels, have strategies and plans to deal with such situations. Such a strategy and plan is a type of insurance against any undesirable situation that may evolve. This may be likened to insuring your house against fire damage. In other words, you need to insure yourself against 'black swans', that is, unpredictable events that can have potentially severe consequences. The severe impact of such an event can ruin people's lives and destroy organizations and societies. One does not have to imagine a disaster film to understand that it is those who have developed their systems for such emerging events who will be best prepared to deal with them. 'Plan or be planned for', said Russel Ackoff (1981, 2006). Related to black swan events is the idea that social systems should create their own future by developing scenarios and plans for the different futures that may emerge. It's not that these know better what will appear on the horizon. It is the planning and exercises that make these systems more prepared should the unexpected occur. This is not unlike having fire drills in order to practice how a building can be evacuated in the event of a fire. Similarly, organizations need to have 'fire drills' to cope with any crisis should it occur; such plans and drills make the system ready for whatever may happen. One should not deconstruct emergency preparedness, but rather the opposite, one should strengthen it. In other words, you need to be prepared for the unexpected, what you don't think is likely, but which is absolutely possible. This is not unlike gambling in the lottery in terms of likelihood. You know that the probability of having the winning lottery ticket is near zero, but you also know that it is not impossible.

As the countries and regions of the world become more intertwined, what happens in one place in the world can quickly affect people in other places. For instance, many scientists believe that Wuhan market in China was the epicentre of the start of the COVID-19 pandemic, which then spread rapidly to other countries around the globe. Innovations, and not least innovations that are systemically linked, can also spread very quickly. Therefore, a technological innovation that is developed in the US can, with a few keystrokes, spread very quickly to the rest of the world.

When such strategies to safeguard against black swans are to be designed, Weick's (1979) 'future perfect thinking' can be useful. What we should be aware of is that 'rules and routines' tend to be preferred over 'visions'. The reason is that rules and routines are easier to deal with than visions. The rules and routines also have legitimacy through their history (Kanter, 1983: 304). In order for the vision to influence rules and routines, the entire pattern, in which these rules and routines are a part of, needs to be revealed. This means that the visionary leader must: 'operate at levels of patterns of changes as well as events (Senge, 1990: 355). This idea also agrees with systemic thinking (Bateson, 1972; Weick, 1979). Systemic thinking informs us that if we want to make changes in organizational systems, we need to change the relationships in these systems. Another lesson from systemic thinking is that if you want to achieve qualitative change processes, you should introduce less central control, less central planning and less central power in the system. The rationale is that the complexity created by change processes requires self-organizing units.

Another lesson from systemic thinking is that if we think in terms of circular processes, then changes in one place in a circular loop can lead to major changes elsewhere in an organization. This means in practice that before we start developing plans to meet the unexpected, we must first examine the relationships and the pattern that these processes are part of. In this way, we will have a greater opportunity to understand and explain how changes in one place in an overall process can affect other elements in other places where the process makes an impact. In circular loops, we should always look for processes and patterns. 'Pattern' is here understood as a system of interwoven elements, which repeats itself over time.

'Future perfect thinking' can be defined as thinking in the future perfect tense (Weick, 1979), which is imagining the future as if it has already occurred. In other words, you visualize what you want to achieve before you start a process. When you have concretely visualized the end result you want, you imagine that you have reached the goal, look back and ask the following two questions:

1. What obstacles did I encounter on the way to the goal?
2. What and who helped me reach my goal?

The purpose of these two questions is to clear the path that must be taken to reach the goal.

Case letter 3: Expectations are fuelled by enthusiasm

We are trained to think about goals and less used to thinking about direction. However, it is rarely the case that the course between two points, where you are and where you are going, is a straight line. If it is a straight line, then it only exists on a map. Mistaking the map for the terrain is something most of us do very often. We confuse our mental models and our perspective with practice and therefore try to make our mental models fit reality. Our 'map' can also be understood as our imagination. Our ideas are made up of our experiences, routines and habits.

We have a strong resistance to changing our thinking and practices. In other words, we are not willing to change our 'ship's' course, even when a storm is looming on the horizon. Some will oppose any change, while others often react with apathy. However, if the new course is adequately staked out, it will be easier to seize the golden opportunity to change our ideas and attitudes. One way to change our attitudes is to change the way we think, speak and act.

There are four processes, which we describe below, that will help to influence changes in attitude. The four processes are:

- Strive to do what others say is impossible
- Nothing is too difficult that it can't be done
- Eliminate habits that do not lead to your goal
- Adapt continuously to changes

Strive to do what others say is impossible

When you want to do the impossible, the first thing you have to do is prioritize your work tasks. A simple technique is to use a pen and 'post it' notes to make a list of the tasks. Thereafter, you have to keep strictly to the sequence of prioritized tasks. Furthermore, you should organize your workdays so that they are divided into sequences according to the prioritized tasks. They are then assessed the next day. It is equally important to delegate to others what is not so urgent, or less important, as well as what is important but not urgent. In this way, you will have more time at your disposal for the priority tasks, those that are urgent and important. However, you should stop doing a task, when you do not feel on top of it, not after you've given your everything. During the work day, other small tasks will appear. Write them down immediately and work on them for a short period at the end of each priority sequence. It is also important to set a cut-off time for each task, otherwise one priority may 'swallow up' the time designated for the next task.

If you let other people's ideas about what is possible and what is impossible dominate your thinking, then you will literally be chained to mediocrity in every context, because there are always some people who have restrictive ideas about what is possible.

Nothing is too difficult that it can't be done

One technique for changing an action pattern is to change some habits, because if you change habits, the action pattern changes itself over time. If, on the other hand, you want to change others, the best way to achieve this is to change your own reaction to their behaviour.

It is through your ability to imagine that you can push yourself further and further so as to increase your scope of opportunities. However, this presupposes that you're willing to lose your footing, i.e., step away from that which is safe and secure. Changing habits and behaving 'as if' it is possible increases the potential for learning and promotes the dynamics of change.

The good news is that you can imagine how you can meet different situations and thus increase your range of opportunities. Sportsmen and women who perform at the highest level adopt this approach.

You don't necessarily need to learn through your own or other people's experiences. You can change your thinking about the possible situations that may arise, and your scope of opportunities will increase. By changing the way you think, for example by simulating different situations in your own 'inner theatre', you can therefore become more flexible in the specific situation that arises. You increase your ability to cope, change yourself and thus the attitudes of others.

An important point here is that what you achieve is a result of how you think about yourself. You start by becoming who you want to be by imagining that you are where you want to be. To be able to do this, you have to lift your gaze above the horizon. Your diadem of thoughts is of course the actions they result in, so one must be willing to sacrifice something.

Eliminate habits that do not lead to your goal

Inspiring yourself with fantasies and mental images can be a help to get you out of your comfort zone, because the purpose of habit is, amongst other things, to prevent you from moving into the unknown and unsafe areas where stress, unpleasant surprises and pain may occur.

Just as the oak slumbers in the acorn, change slumbers in our thoughts and ideas. However, resistance to change is built into our mental constitution and acts as a safeguard against unpleasant surprises. By thinking about what can happen in different situations, stress is reduced and we are able to develop momentum for change to take place and help us to prepare for change. The strongest power we possess is our thinking. The more often we think over imagined situations, the greater the impact our thinking will have on the development of situations. This expectation mechanism is analogous to 'self-fulfilling prophecy'.

If we let our thinking run free, the brain will expend too much energy finding out how to avoid change. The same mechanisms seem to come into play if we have achieved success. We try to repeat what led us to success by using the same action strategies, without reflecting on the fact that the world around us has changed.

Reactive thinking is based on past experiences, which we then project to deal with a current situation. Proactive thinking starts from what is possible and then moves backwards in time, from the future to the present, to examine what promotes and what hinders us from achieving what is possible. Inactive thinking is mainly based on what is happening around us here and now, and then we try to adapt to this as best we can. Interactive thinking[16] connects all three modes. Systemic thinking is to model the pattern that people with obvious success in the area we want to improve have used to achieve their results.

Adapt continuously to changes

We can vary our thinking when setting a course and make changes to avoid the 'hidden rocks in the sea'. The point of continuously adjusting our course is to avoid the snags, which will always appear. For that we need alternative ways of thinking, because how we think is often more important than what we are thinking about. The different ways of thinking require practice to be able to be applied in specific situations. The best practice in this context is to run through different situations over and over again in our 'inner mental theatre', where we apply the different ways of thinking to exactly the same situations. What we then do is to develop a greater scope of opportunities. This expectation mechanism can then be used to correct our course.

The more often one alternates between different ways of thinking, the greater the probability of success when the complexity is great. Indeed, the more often you use customary modes of thinking to solve new challenges, the greater the likelihood that you will lose opportunities for change, when the complexity of the world around you increases. Actively using different ways of thinking engages curiosity, increases the scope of opportunity and promotes creativity.

There are countless examples of new ideas that have become great successes, precisely as a result of people being willing to look at a failed project as the starting point for an innovation. The point is that you need to adapt no matter how a project develops.

What we have tried to emphasize here can be formulated in one sentence: 'Our way of thinking influences our actions'. It is therefore crucial how one thinks. We are trained to stay focused on what we are thinking about, and less on how we are actually thinking. Using the power of thinking enables us to continuously adapt to change.

Expectations are fuelled by enthusiasm. When we have less enthusiasm, we lose interest in making the extra effort that is needed to reach the goal. Whether you win or lose depends on the effort you make and no battle can be won without genuine enthusiasm.

Description related to black swan events

Waves of black swan events or global policy crises can be considered on many levels and from different aspects. We choose here not to focus on the philosophical and epistemological approaches, not because it is uninteresting, but because it is outside the scope of this book. Our focus here is black swan events related to the workplaces of the future. Therefore, we will approach this issue from a more practical level. The particular aspects we have chosen to consider in relation to the issue we are examining are: Systemic innovations, increasing complexity, co-creation, contextual competence and intelligent robots.

There will be limits to how much people, organizations and countries can adapt to the many changes that will emerge during the approaching Fourth Industrial Revolution. The shock caused by such radical changes can literally paralyse people and make them unable to act rationally in relation to the many demands the changes will make. Psychologically, the 'shocks' caused by these changes can lead to traumas. Trauma can be passed down from one generation to the next, so-called intergenerational trauma, because social learning leads to these psychological 'wounds' of life being inherited. In this context, it can be interesting to consider the traumas that the shock of change leads to. From such a perspective, the Marxist class struggle is not something that the individual or his family alone is exposed to, but a generational class struggle, between capital and workers, passed down from generation to generation. Between these two classes is the middle class. The classes are not strictly rigid, as there will always be people who manage to adapt, and move up a class, so-called intergenerational mobility.

In this context, the systemic innovations represent the black swan events we have referred to above. These systemic innovations will emerge as waves of change resulting in 'shocks' to people, organizations and social systems. At the same time, the innovations, especially those related to artificial intelligence, will also lead to growing geopolitical tensions, particularly between China and the US (Johannessen, 2021a, 2023b, 2023c).

'Adaptability' might sound like Darwin's theory of 'survival of the fittest', that is, adapting to changes in the environment. However, the opposite of adapting is not necessarily not adapting. The opposite of adapting may be viewed as creating something new or creating one's own future. Creating one's own future is here understood as the opposite of adapting to what others have created. This will lead to increasing complexity and promotes co-creation, as a necessary, if not sufficient, prerequisite for developing contextual competence that triggers the creative new. In the upcoming Fourth Industrial Revolution, intelligent robots based on artificial intelligence, that are the result of contextual competence, will have been further developed. In this context, it is not difficult to envision a rush of black swans impacting social systems with waves of radical changes.

The adaptability of the individual is not infinite; it is severely limited. For the individual, it is limited by biological conditions. At the same time, the ability to adapt is limited by social opportunities. These social opportunities to survive through adapting can be limited when there is total social chaos. When chaos occurs in social systems, adaptation is not necessarily the best strategy, but rather creating one's own future through structures that establish stability.

Adaptation is possibly the most dominant social mechanism related to changes in the environment. The point in this context, however, is that when we continuously choose to adapt, we are not necessarily only exposed to erratic behaviour, but also become a type of 'psychological amoeba', a person with no individual qualities, other than the ability and willingness to adapt to the demands of others. You can imagine the demands related to change like a barrel you fill with water. As you change through critical periods in your life, you fill up this barrel. When the barrel is full, the water starts to flow over the top edges and out into the areas around the barrel. It is at this point, the psychological consequences of this adaptation strategy become visible, not only to oneself but also to the outside world. If we extend this analogy, then some changes in life fill the barrel more quickly, while other changes hardly add any water at all. You'll never know which changes will be the droplet that will result in the barrel overflowing. What we do know, however, is that when it 'overflows', the psychological and social consequences can be destructive. Having to continually adapt to radical changes over time can easily lead to both mental and physical illnesses (Watzlawick et al., 2011: 31–75). For businesses, this can lead to a lack of ability to innovate and bankruptcy even if productivity is very high. The connections between change stress, adaptation strategy and various forms of illness for people, and bankruptcy at organizational

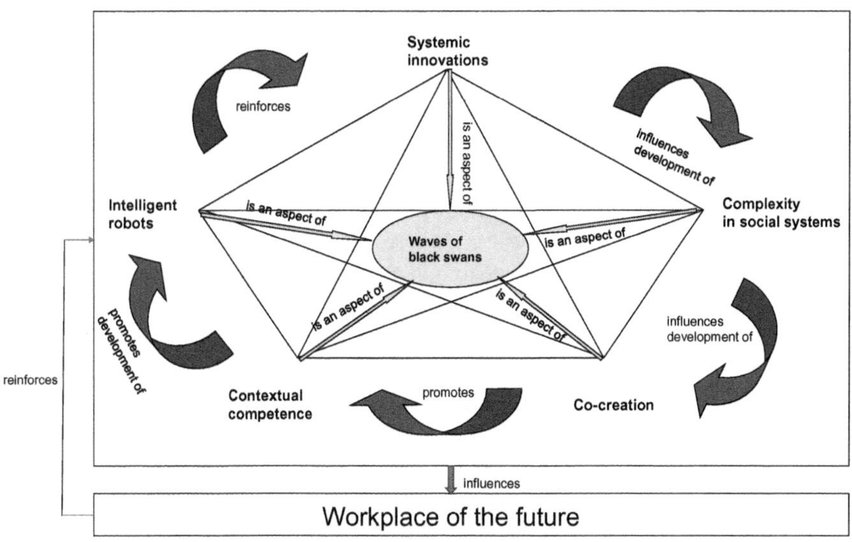

Figure 6.1 Waves of black swan events: A conceptual model.

level can have catastrophic consequences. It is here that we can say that black swan events can result in people and organizations choosing adaptation strategies that do not promote the desired goal.

Based on the above description, we have developed a conceptual model that indicates some aspects of black swan events that will emerge in the innovation economy (Figure 6.1).

Analysis related to waves of black swan events

In the following analysis, we will give a detailed description of the five aspects in Figure 6.1.

Systemic innovations

The consequences for the labour market of systemic innovations are almost impossible to imagine. It may be true that we do not have knowledge of an unknown future. On the other hand, we can use our mental capacities to develop conceptual models. These models can then act as maps for our future behaviour and actions. Once we have developed these maps, they can act as social mechanisms to change the workplaces of the future. In this context, one might say that the map will not necessarily represent the terrain that emerges in the future. We are used to thinking that we draw maps that represent an existing terrain. However, regarding systemic innovations, it may be useful to draw a map that is capable of changing 'the terrain'. The process is as follows:

1. First, we consider how we want our workplaces to develop in a desired future.
2. We then develop conceptual models for this process.
3. We then use the systemic innovations that will emerge in the innovation economy as input factors to prepare workplace development readiness.

It is precisely the concept of development readiness in point three above, which is important here. Development readiness is something completely different from adaptability. Adaptability is something Darwin used to explain and understand the evolution of nature. In social systems where people and technology work side by side, we believe it is not appropriate to transfer knowledge from nature's evolutionary development to how social systems are developed. It is not adaptation, neither passive nor active, which is the appropriate strategy for understanding and explaining or changing the development of social systems, but readiness for development. This concept is creative and innovative. This can be related to Ackoff's (1981, 1999) concepts of 'creating the future' and 'plan or be planned for'. We can also relate 'readiness to change' to Weick's (1979) concept of 'future perfect thinking'.[17] Both Ackoff's and Weick's concepts can be used to create the map that will change the future terrain, that is, the workplaces of the future.

If we create a map that changes the terrain, we will have the opportunity to limit or even eliminate the robot shock that will affect our social systems and our future workplaces. It is in this context that we can say that what we are doing is creating the social reality to which we must then adapt. The sequence is thus as follows. First, we prepare the social system for change by techniques such as 'create the corporation future' and 'future perfect thinking'. Then the social reality will unfold that we ourselves have helped to create. When this social reality emerges, we adapt to this reality. In this way, evolutionary events in social systems become distinctly different from evolutionary development in nature. In nature, it is the random process of natural selection that enables species to adapt to their environments. In social systems, it is conscious creative and innovative strategies that function as the selection mechanism. The distinction between evolutionary processes in nature and in social systems is found in the relationship between random selection and deliberate selection. Social systems are not left to chance; they are developed by intentional, desired and conscious strategies. Weick expresses this idea in the following way: 'How can I know what I think until I see what I say' (1979). The point of this statement is that one must develop some conceptual models that create a map of the desired terrain. Then the social reality develops on the basis of these maps which change the social reality. The idea here is that if we do this, we will be able to control the systemic innovations within a certain cybernetic control area. By this, we mean that we will not have full control over how the systemic innovations affect our future workplaces. On the other hand, we will have some control over the consequences within upper and lower control limits in the cybernetic control model (Johannessen, 2020).

The main message in this section on systemic innovations is related to the fact that when we work together in groups, teams and social systems, how we create our future is not something random like a lottery game but a co-creative process. Random selection is something that applies to the evolution of nature. However, people have a state of consciousness and can decide their own futures so that developments are not left up to chance as in nature. It is precisely this quality that enables us to create the maps that shape the terrain of the future, that is, our workplaces.

Complexity

Systemic innovations increase complexity, because the changes come faster, and they are interconnected and difficult to obtain an overview of. When leading in such situations, a simple piece of advice can be to not try to control what happens, because this will mainly lead to increasing complexity. The strategy for dealing with complexity is to increase the degree of simplicity. For the individual employee at a workplace that is exposed to increasing complexity, a simple piece of advice is to focus on what you like doing, what you master and what you are good at. Trying to satisfy the demands of a leader who attempts to cope with increasing complexity by increasing control will not, as a general rule, be an appropriate strategy.

In analogy to the title of Schumacher's book, *Small is Beautiful* (1993), one might say 'make simplicity out of complexity'. In workplaces, most leaders will deal with complexity by making the working situation more complex. Such thinking can be related to the idea of 'requisite variety' (Ashby, 1970, 1981), where a system, in order to fulfil its purpose and survive, needs to be capable of a greater variety of responses than the variety of factors that exist in its environment. In most cases, such thinking will be appropriate. However, regarding systemic innovations, the application of 'requisite variety' will only lead to greater complexity and less control. Systemic innovations may be compared to ecological systems. When you intervene in ecological systems, you disturb relationships that should not be disturbed. When this happens in ecological systems, more relationships occur, thus increasing complexity. This is exactly what happens regarding artificial intelligence and their connections to networks. When an attempt is made to shut down networks, artificial intelligence, like an ecological system, will find new connections. These new connections then create increasing complexity in their consequences. What can be said is that the attempts to control only lead to the need for more control. The basic idea is that in complex ecological systems, control is lost when more control is introduced. It is against this background that one can say that complexity is best handled by introducing simplicity.

Simplicity must not be confused with passivity. In our context, simplicity is related to commitment and active participation in work situations. Workplaces of the future will most probably extensively utilize AI technology. When leading such a workplace, it will, in many cases, be most opportune to focus on the front line. In other words, the employees who are directly involved with doing what the system is designed to do. The leader who steps back a little, and who delegates decision-making to front line employees, will lead by embracing simplicity. The leader who attempts to increase control in such change situations will find that the system gets more and more out of control. The classic response in such situations is to increase more control. The consequence may then be that the system collapses, because too many resources go to control functions and not what the system is designed to do.

Disorder and disarray do not lead to complexity but to chaos. Chaos in social systems exists when there are no predictable patterns and when there is little or no interconnection between the elements of the social system. To deal with chaotic systems, the technique is to add links between the elements where no links or relationships exist. It is when we create the connections that do not exist in chaotic systems that we make these systems complex, and thus manageable. They are manageable precisely because we use simplicity to intervene with these complex systems.

Acting on the basis of simplicity in a leadership position when black swan events enter workplaces or more concretely systemic innovations become part of everyday life is based on knowledge of how complexity is handled in social systems. It is neither passivity nor the absence of competence that characterizes simplicity but a large degree of competence about complex social systems.

Co-creation

We know that systemic innovations lead to increasing complexity. We also know that we can reduce complexity by working together but most likely co-creating will give us even greater control over complexity. When we talk about co-creation here, we mean not only that people work together to co-create but also that we turn to nature to a greater extent to find forms, patterns and objects that we can use to generate innovations. The tendency to learn from and mimic the strategies found in nature in order to create innovations has gained much ground from the 2000s onwards (Harman, 2013: 22); in other words, biomimicry and nature-inspired innovation. Nature-inspired innovations take as their starting point nature's processes that have been developed and adapted to changing environments over millions of years. Nature-inspired innovations can be found in many different fields, such as in the pharmaceutical and medical fields (Benyus, 2002). A great number of today's pharmaceutical products used in modern medicine are derived from plants. Using industrial processes, active ingredients in medicinal plants are isolated and synthesized, enabling the production of pharmaceutical products, '(everything) from stimulants to sedatives, painkillers to chemotherapy, detoxicants to antioxidants, and cardiotonics to antidepressants' (Harman, 2013: 29). The point of bringing this up here is to highlight that co-creation to a large extent concerns knowledge development and collaboration, as well as investigation of processes in nature. However, our focus here is on the workplace of the future. The question that arises in this connection is the following: What can nature teach us about how to organize our work, when changes occur faster than we can adapt to them?

We will not provide in-depth answers to the question here but only suggest what organizational innovations inspired by nature might look like. One way to transfer this type of knowledge could be to study specific systems in nature. For instance, one could examine the social organization of various species of insects and animals, such as ants, flocks of birds, schools of fish and so on.

If we start by looking at the organization of a bee colony, then it is clear that bees have a type of functional organization, i.e., the various categories of bees have specific functions.[18] For example, young worker bees clean the hive. They progress to other tasks as they become older, such as receiving nectar and pollen from foragers, and guarding the hive. Worker bees cooperate to find food and use a pattern of 'dancing' (known as the waggle dance) to communicate information regarding resources with each other. In other words, bees use their communicative dances to tell other bees where to find food. Bees dance in circles (a round dance) when they want to show that a source of nectar is less than 50 metres away from the hive. They perform a figure-of-eight dance (a waggle dance) if the food is more than 50 metres away. The faster the bees dance, the more distant the food source. In order to direct their fellow bees to places with the most sugary nectar, the bees will reject the bees returning to the hive with the least sugary nectar (a type of quality control). In other words, these bees are not allowed to dance. The bees that arrive with the most sugary nectar are encouraged by the 'transport bees' to dance so that all the other bees can follow them. In this way, the colony optimizes its nectar-gathering activities (Seeley, 2019).

With these simple rules, the bees in the bee colony communicate information about the quality and quantity of the nectar and the distance to the nectar. The distance is important because it says something about the productivity of collecting the nectar, because bees expend energy by flying, so long distances are not as productive as short distances.

In our context, there is one aspect of the organization of bee colonies that is of interest. The bees operate with flexible working roles with varied work tasks. In the bee colony, the bees have specialized roles, but they can change roles, and are therefore multifunctional. For instance, bees can change their roles from cleaning and guarding the hive to collecting nectar and so on. If we transfer this idea to the health sector, one could imagine that the fixed functional areas in hospitals could become more flexible. In practice, one could imagine that bioengineers and radiographers could take over some of the medical specialists' routine tasks if the situation required this. In this way, the medical specialists would be able to free up some of their time for more specialized tasks.

If we transfer this idea to the workplace of the future, one could imagine that people's competences were structured according to a T-structure, i.e., people's qualifications could include both in-depth expertise and broader knowledge. In this way, the various health professionals working in a hospital could have supplementary skills in addition to their core expertise, so they could be more flexible regarding work tasks. In such a way, co-creation and contextual competence would gain a new meaning in relation to organization when changes occur very quickly.

Contextual competence

Contextual competence is understood here as a link between part and whole, where competences that were previously disconnected are now linked, even though each type of competence is distinct. Contextual competence can be metaphorically understood as a synthetic brain where learning algorithms control competence development. Such a synthetic brain can be said to exist today in intelligent robots. Contextual competence is thus closely related to artificial intelligence and intelligent robots, which we will elaborate on in the next section.

In concrete terms, contextual competence for the question we are investigating here means that we have the opportunity to adapt to changes in a qualitatively better way than we would be able to do without such competence.

Contextual competence can possibly be best understood in relation to a 'cybernetic brain' (Pickering, 2010). Cybernetics is not the same as automation. Cybernetics in an organizational context can be defined as management, control and communication (Wiener, 1948). In an organizational context where one tries to understand an organization's culture, a cybernetic perspective will have as its starting point that there are many different cultures within an organization, so it will not be possible to identify a uniform culture but rather many different aspects which together constitute what can be called the organization's culture (Pask, 1992: 11).

Contextual competence from a cybernetic perspective can be understood as an examination of what is disconnected; that is, competences that are not in contact with each other with the intention of achieving a purpose. Our aim here is to examine the increasing rate of change which will result from the future increased application of AI technology in our workplaces. In its simplest form, cybernetics is an investigation of interconnected feedback systems where information is sent and received over a network (Ashby, 1961, 1970).

Stafford Beer's cybernetic system can give us insight into how waves of changes can be stabilized by using a special organizational design (VSM). Beer's book, *The Brain of the Firm* (1981) describes how an organization can be modelled analogous to a human brain. We will briefly refer to this design in relation to dealing with changes that occur faster than we can adapt to. In other words, when our workplaces are significantly affected when black swan events appear. One would think that when changes flood our workplaces, they would be negatively affected, become non-functional and lose their effectiveness. With an overarching design, such as Beer's viable system model (VSM) (1995), the organizational structure can retain its shape, even when changes radically impact people's skills.

Beer's organizational model consists of five distinct but interconnected systems. System 1 is what the overall system is designed to do. There can be many system 1 units. For instance, in the university sector, these units may be categorized into: Teaching, research and dissemination to the field of practice. However, these system units must have a controlling and managing system that allocates resources. System 2 acts as the controlling system. This system has developed some clear goals for the system 1 units and ensures that they meet certain goals. If the goals are not achieved, then it is system 3, the 'here-and-now' management, which intervenes. This intervention can be oriented towards information and communication or towards resource allocations to enable the individual system and units to achieve goals. System 4 may be termed 'an eye to the future'. This system ensures that the organization is constantly updated regarding changes in the external world and also helps to create the organization's external environment. In this system, a conflict may arise internally between those who focus on adapting and those who are concerned with creating the organization's future through various creative processes aimed at developing innovations. A conflict can also arise between system 4 and system 3, because it is often the case in organizations that those who are concerned with the 'here-and-now' that are allocated resources, because they have to deal with urgent and pressing matters. To mitigate this potential conflict internally in system 4 and between system 4 and system 3, the Viable System Model includes a system 5. System 5 operates as the overall decision-making system that balances the conflicts, especially between system 4 and system 3 and ensures that the organization maintains sustainability in relation to what the system is designed to do.

Beer's VSM model is a resilient model, and manages to remain stable, because the individual systems change as the environment changes. This may be compared to the tightrope walker who constantly has to change the position of his/her arms and legs to remain 'stable' on the tightrope, so as to reach the other side. In other words, the prerequisite for stability is change, and stability is maintained when changes occur in the various systems.

Intelligent robots

In the future, applications of artificial intelligence technology will automate many of the work activities (Johannessen, 2020, 2020a). At the same time as this automation process progresses, the education system and the attainment of new skills will come under pressure. This will most probably eventually result in the development of skills that are compatible with the new technology (Johannessen, 2020b).

A distinctive feature of the utilization of artificial intelligence in social systems is the increased speed of information and communication processes (Johannessen, 2020c, 2020d). This not only affects automation and innovation processes but also puts ethical questions on the agenda (Johannessen, 2021).

The above leads to the question that several commentators have asked: 'Can computers think?'[19] If we believe that applications of artificial intelligence can reflect on facts, rules and logical patterns, then we might answer that it is probable that artificial intelligence can learn 'to think'. If, on the other hand, we believe that there is a clear distinction between rationality and human reasoning, where human reasoning includes emotions, social relationships and consciousness, then it is highly unlikely that artificial intelligence (AI) can ever learn 'to think' like humans (Searle, 1995). However, if we consider AI integrated into robotics, so-called intelligent robots, then although these robots cannot reflect like humans about emotions and social relationships, they will nevertheless influence the thinking of those people who are affected by these new robots. In other words, it is how these robots will concretely affect our workplaces that most people will be interested in, not the fact that the robots lack human consciousness. Although the question of 'consciousness' may be of interest from a philosophical perspective, it will not be of interest to those people whose jobs are negatively affected by the introduction of robots in the workplace. Despite the fact that the robots will have a 'limited way of thinking', they will have a unique ability to imitate human thinking with regard to most work processes (Johannessen, 2022a, 2022b).

The vast majority will be concerned with how the utilization of intelligent robots in the workplace will affect their working life. This development will very likely come as a shock wave to many. If people and social systems just wait for the shock wave to hit them, before they act, then the problems will probably be too large to deal with. It is then that we will see people, organizations and social systems being impacted by various types of crises. It is in this situation that governments and others will have to take action to pre-empt the shock waves. This can be implemented at the governmental level by developing various strategies that will enable people, organizations and society to cope with the major changes that the shock will result in.

It is highly likely that it will be the rational, logical and linear work processes that will be affected first. However, we should note that high-tech automation will first result in productivity falling before it rises. It is during this time lag that businesses will be reluctant to be 'the first movers' when more AI applications are introduced into workplaces. However, it is precisely during this period that people and businesses should invest in resources to prepare for the upcoming shock waves. In other words, it will take time before the utilization of the new technology reaches optimal levels. Therefore, initially, we will not witness dramatic changes in workplaces. However, it will be important for people and organizations to use this time lag to adapt to increased AI automation. If people and organizations are slow in making this transformation, then it is highly probable that the shock waves will lead to small and large crises for people, businesses and society.

Theoretical reflections related to waves of black swan events

Today, we often read in the newspapers and academic journals that the utilization of artificial intelligence (AI) is resulting in many jobs being at high risk due to automation. Moreover, many of these are highly skilled occupations, such as the translation profession. Changes in this section of the labour market is resulting in many translators having to carry out different work functions than before. For example, many translators are now post-editing machine translation; in other words, they are becoming proofreaders in one sense. In this context, one can say that in many cases intelligent machines are changing work functions but not replacing them. Furthermore, machine translation is being continuously improved because translators send the machines feedback. In its consequences, this improvement in machine translation leads to less demand for translators in the labour market.

However, it is the extent of AI applications in workplaces that will determine the extent of automation. In other words, the link between AI applications and systemic innovations will result in many jobs being at risk due to automation. Of course, this development need not be a negative one. To what extent it will be negative or positive will depend on which governmental policies are implemented, when people need to find other types of work. If the political ideology promotes productivity, while also ensuring that people's welfare takes precedence over profit, then the increased use of systemic innovations and AI automation will most probably be mainly positive for people and social systems. If, on the other hand, the political ideology focuses on profit over people's welfare, then increasing automation could very well lead to a deterioration in people's well-being. In other words, when black swan events make their appearance, the consequences for most people will be determined by which political policies are implemented.

An interesting insight, originating from the ancient Greek philosophers, Aristotle and Plato, is that you don't write to show what you know, but you write because you don't know. In other words, questions are more important than answers because questions seek to understand something, while answers are only temporary and often need to be re-evaluated over time. This insight is also related to the ideas proposed by Umberto Eco in his various works (1992, 1994, 2000, 2002, 2003). The point of this insight in this context is that if someone asks why you write about a subject that you know nothing about, the answer is simple. If I had known anything about the question I am researching, I could have just taken a book from the bookshelf that deals with this very question and shown it to you. So what does this insight have to do with the question being investigated here? Regarding focusing on what you know, instead of what you don't know, Taleb says: '--by focusing on the known is a human bias that extends to our mental operations' (2010: 1). It is precisely here that this insight can be related to the black swans that Taleb writes about and the black swan events that we are examining in this chapter. The point of black swans is that, first, there is a distinction that is crucial, so crucial that it is the difference that really makes a difference, which Bateson (1972) often refers to. This difference is precisely expressed between what we know and what we do not know. We do not conduct research in those areas where we have knowledge but in those areas we know nothing about.

We know that the further development and applications of artificial intelligence will increase the rate of change in workplaces and society so that people, organizations and social systems will be strongly affected. We also know that this will greatly impact our workplaces (Susskind, 2020). What we know little about, however, is precisely how, and what happens when people are unable to adapt to the many changes that will occur, when systemic innovations and AI applications are increasingly used in workplaces. This is precisely why we reflect here on some of the theoretical relationships and develop propositions in order to illuminate these relationships.

The propositions developed below are research tools that can be used to reveal some, but not all, of the possible relationships that follow from the question being investigated in this chapter.

The idea of this theoretical reflection is very simple. It is the link between the degree of application of intelligent robots and systemic innovations, which gives rise to four connected but distinct hypotheses.

Hypothesis 1: If the degree of application of systemic innovations is high, at the same time as the degree of application of intelligent robots is high, then the probability is high that we will see waves of black swan events entering the labour market.

Hypothesis 2: If the degree of application of systemic innovations is high, while the degree of application of intelligent robots is low, then the probability is high that we will see the emergence of chaotic systems.

Hypothesis 3: If the degree of application of systemic innovations is small, while the degree of application of intelligent robots is high, then the probability is high that we will see robust systems in the labour market.

Hypothesis 4: If the degree of application of systemic innovations is small, at the same time as the degree of application of intelligent robots is small, then the probability is high that we will see vulnerable social systems and a weak labour market.

Based on the above reflection, we have developed a typology for the relationship between systemic innovations and intelligent robots (Figure 6.2).

Practical utility related to waves of black swan events

Technological development promotes increasing complexity in most social systems (Zuboff, 2019: 376–398). This increasing complexity compels a greater degree of collaboration not only between areas of knowledge but also across national borders (Sennett, 2013; Davenport, 2005, 2019). It is in this process that knowledge

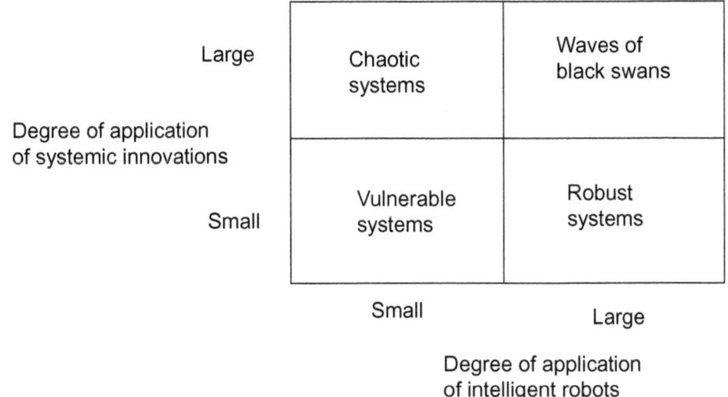

Figure 6.2 The relationship between systemic innovations and intelligent robots: A typology.

boundaries are traversed and a greater degree of contextual competence becomes useful for co-creating innovations, for example, within the field of technology (Pfeffer & Sutton, 1999; Johannessen, 2023). However, these creative and innovative processes have an unintended effect. They create systemic innovations that impact the market as waves of black swan events. These black swans or systemic innovations result in increasing the complexity even more.

The increased use of artificial intelligence (AI) and robotics will lead to increased digitization. The further development of robotics and AI will also lead to more surveillance and monitoring of people in the West (Zuboff, 2019) and in China (Johannessen, 2021a, 2023b). At the same time, there are also tendencies that indicate that the new technology will reinforce the development of a new form of capitalism, feudal capitalism, both in the West and China (Johannessen, 2023a). This development will also lead to increased tensions between China and the US and reduced trade and technological cooperation. In other words, an increase in de-globalization between West and East (Johannessen, 2023c).

What is also becoming evident in relation to the above developments is the fact that data is becoming the 'new oil', so to speak. The background for this assertion is straightforward. In the agricultural society, it was the ownership of land that determined power relations. In the industrial society, it was the ownership of capital and capital goods that determined power relations. In the innovation economy, it is data that will be the central factor that determines power relations. We can take an example. China monitors its citizens through Internet surveillance, camera surveillance and other digital technologies. This enables the government authorities to collect enormous amounts of data. When this data is analysed, it provides the basis for transformation into information and knowledge. This knowledge can then be used to develop theories and concrete policy decisions to control people's behaviour. This is also occurring in many other countries. Moreover, data analytic companies have harvested data illegally and used the data to provide analytical assistance in democratic elections. The dance around the golden calf has today been replaced by the dance around data. Access to data and the development and applications of AI will lead to increased changes in working life, which people will have difficulty in adapting to. The increasing amounts of data and the consequent increase in information to which people will be exposed may very well have negative psychological consequences.

In a situation characterized by information input overload the following may occur (Miller, 1978: 123):

1. Designated tasks and responsibilities are left undone.
2. Errors are made.
3. Queues of information occur.
4. Information is filtered out that should have been included.
5. Abstract formulations are made when they should have been specific.
6. Communication channels are overloaded, creating stress and tension in the system.
7. Complex situations are shunned.
8. Information is lumped together for processing.

Each of the above eight points may result in a decrease in efficiency when the system is exposed to information input overload.

These eight factors may also lead to businesses investing further in artificial intelligence and intelligent robots in order to increase productivity. In this context, it is perhaps appropriate to refer to the subtitle of Paul Watzlawick's book (1988) *Ultrasolutions:* 'How to fail most successfully'. In other words, businesses that fail because when solving problems, they adopt more of the same strategies that only make matters worse.

Zuboff argues that we are moving towards an age of surveillance capitalism. It may also be proposed that we are moving towards a society with even more control, a feudal state governed by a form of feudal capitalism, which the author of this chapter and book argues for (Johannessen, 2023a). This would mean that the privileges of a small minority would continue to be strengthened. In practice, this would mean that this structure can only be maintained with increasing pressure on the development of innovations. The rationale is that it is these innovations that will make it possible to increase automation and to reduce the dependence on human labour for production. However, there is a part of this innovation process that is dependent on human labour, and that is the production of knowledge. This knowledge is precisely the input factor for the innovation processes. Against this background, in the future, we will see a greater demand for knowledge and innovation workers. It is this innovation focus that will strengthen the development of systemic innovations, which in turn will lead to increased complexity. It should be emphasized that this development is not dependent on the emergence of surveillance capitalism or feudal capitalism but an economic policy where economic growth and profit drive economic development. The point of mentioning these two forms of capitalism is to emphasize that any development, regardless of ideology, will be based on economic growth, which will promote the development of innovations and thus increase the complexity of most social systems.

The automation resulting from the new technology will, as indicated above, promote a focus on knowledge production. There is, however, an important point about this knowledge production in the innovation economy, which is largely different from before. In the past, knowledge and innovation could be developed through analysis and in-depth knowledge. This will not disappear in the future but a significant new element will be added. As more and more systems, not only knowledge systems but also other systems at a different level, are in contact with each other, the development of knowledge will take a slightly different direction. This direction will be a link between in-depth knowledge and broader knowledge, which is then put into context to generate innovations. This initially sounds a little abstract and needs more explanation. We can look at a practical example to show what is meant by the emergent new domain of knowledge that is developing, which we term contextual competence. In the beginning, when computers were first developed, it was largely technological expertise that was important. Of course, this is not entirely true, because there were many researchers from different fields that were involved, which is evident from Ashby's (1961, 1970, 1981), Wiener's (1948) and Pickering's (2010) works, to mention a few. Our point here is that in the innovation economy there has been more focus on merging knowledge domains, domains that were previously not in contact with each other. Contextual competence leads to systemic innovations. There are many examples of systemic innovations that have been created by contextual competence: Biotechnology, nanotechnology, gene-editing, molecular computing, AI protein folding, brain-computer interfaces (BCIs) and artificial intelligence (AI), AI and game development, AI and data analysis, AI body sensors, AI environmental sensors, AI and medicine and so on.

The point here is that this contextual competence requires collaboration across domains of knowledge that were previously separate. This collaboration manifests itself as co-creation and in the generation of innovations. These processes will lead to the further development of intelligent robots. Of course, it is not the case that these intelligent robots will resemble humans. Many of these robots will most likely be nano-sized, such as nanorobots located in the networks people use. Much smaller robots, known as 'nanobots', can be used to clear clogged arteries. In the future, AI robotics can be used to make the traffic in 'smart cities' safer. The point of mentioning this development of AI and robotics in the future is to emphasize that this will impact our future workplaces at all levels. One of the consequences is that the new things that emerge from these various processes will result in the emergence of black swans. These are the unexpected events that suddenly appear and create uncertainty for many people.

Based on the above description, analysis, theoretical reflections and the review of the utility value, we have developed a Boudon-Coleman diagram showing the relationships between waves of black swans and changes in future working life (Figure 6.3).

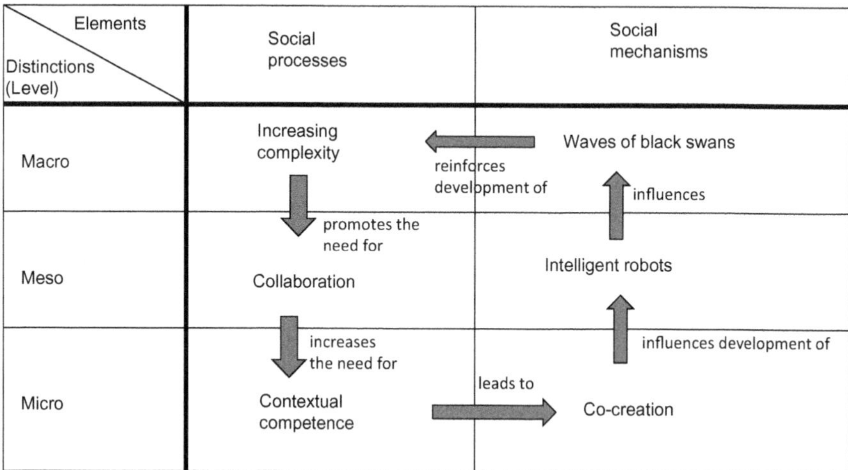

Figure 6.3 The relationships between waves of black swan events and changes in future working life: A Boudon-Coleman diagram.

Systemic connections related to waves of black swan events

Whether intelligent robots will be viewed as a problem will depend on which perspective one takes. If such robots lead to people losing their jobs, then it will most likely be perceived as a problem for those who become unemployed. For companies, however, this may be viewed as a positive development for many reasons. Productivity will rise, labour costs will fall and the dependence on workers who demand a larger share of a company's profits will be reduced. In the longer term, it is also conceivable that those who become unemployed will benefit from these intelligent robots, because some will be able to re-train and learn new skills that are more compatible with the new technology. This can lead to them getting jobs that are better paid, so they can secure their future.

Regardless of which perspective one takes, the intelligent robots will lead to greater complexity at all levels. This will apply to people, businesses, society and not least to the geopolitical system.

If you do not find solutions to the problems the intelligent robots will create at all the system levels, then you may risk solving the problem on one level but creating bigger problems on another level. That is why in this section, we are oriented towards systemic connections and not linear causal connections. To solve a problem in systemic contexts, the point of intervention must be different than if the problems are isolated and not connected to other problem areas. It is for this reason that co-creation and contextual competence are crucial if one is to solve systemically connected problems so that the solution of one problem does not create an even larger problem at another level. Somewhat jokingly, as a simple example, one could say that a conflict between employees and a company's owners ceases to be a 'problem' if the company goes bankrupt due to high labour costs. The conflict was 'resolved', but the company and the jobs were destroyed in the process.

One can call solutions to problems that only create bigger problems for 'hubris solutions'.[20] In Greek mythology, hubris refers to excessive pride or dangerous overconfidence, where one believes that everything is possible. One such perspective related to AI technology is when this technology is applied in military systems. There is little point in trying to create a society based on ethical principles and ethical debate if autonomous military systems are also created using AI technology. In other words, such AI-powered autonomous military systems will ultimately do more harm than good (Umbrello, 2022). In his book, *Designed for Death: Controlling Killer Robots* (2022), Steven Umbrello has reflected on how we can control future AI-powered war technology (so-called 'killer robots') in an ethically acceptable way. However, some of us remember the Soviet nuclear false alarm incident in 1983. Despite the stringent regulations the Soviet Union had for launching its nuclear weapons, it was shown that the system could fail. Stanislav Petrov, a lieutenant colonel of the Soviet Air Defence Forces, judged the nuclear early-warning system's report of a missile launch from the US to be a false alarm and disobeyed orders preventing an erroneous retaliatory nuclear attack on the US.[21] In other words, it was only human judgement that prevented this catastrophe. The point is that no matter how good the control is that humans think they have of the artificial intelligence that is integrated in weapon systems and no matter how advanced this technology is, there will always be some weaknesses in this technology that can cause humans to lose control. In other words, if we view these weapon systems from a systemic perspective, there is a connection that is decisive and which we think Umbrello under-communicates in his book.

This simple connection is that although values are important, technology affects these values so that they change when the technology is used. One can say that technology affects values and the changed values in turn affect technology. This is on the same level as saying that the work environment affects performance but that performance in turn can affect the work environment. If you have a model that says that it is the working environment that affects performance, then this model provides guidance on how to improve performance. If, on the other hand, you have a model that says that performance affects the working environment, then you will have completely different strategies for developing the workplace. Transferred to the connection between military technology and values, the same type of consideration can be used as a basis. Therefore, it will be important when reflecting on artificial intelligence in weapon systems and values that you take both models into consideration. It is in this context that one can refer to the late Stephen Hawking and the open letter that he and many other AI experts signed, warning about the uncontrolled development of AI.[22] Their point was, and is, that it is desirable but almost impossible to manage and control artificial intelligence. If this is correct, then one should tread carefully when claiming that artificial intelligence can be controlled through value design and value assessments. In this context, it is perhaps appropriate to refer to the title of Gregory Bateson's book: *Angels Fear*,[23] which refers indirectly to the saying, 'Fools rush in, where angels fear to tread'. In other words, humans are often foolish and reckless and attempt to achieve feats without thorough reflection.

The criticism against Umbrello's reflections is largely related to what he calls fully autonomous air force systems. To a large extent, his arguments are perhaps valid regarding most of the signals that trigger a response from these systems. The point, however, is that there will always be cases that an algorithm designer has not thought through thoroughly and which may trigger responses from these weapon systems that are not desirable. We have seen this in the discussions of the defence system that is designed to defend the US. Many experts, who have insight into these programming techniques, have stated that in some contexts, the weapon systems behind the defence system could be accidently triggered by built-in errors in the programs. When the weapon system is of a type that it has the potential to create catastrophic conditions, then one should think twice.

Umbrello's solution to this possible problem is that the designer and the user need to have extensive communication. This sounds plausible, but the question is not whether this is a necessary condition but whether it is a sufficient one. How can such communication correct any built-in errors and deal with unknown situations that can arise in a conflict situation? If one were to propose a way of reducing any crisis that may arise, then it would have to be that designers, users, experts in ethics and AI experts, joined in such communication. One can also ask what does Umbrello mean by 'users' in this context. Is it the military that will use the systems? Are they the ones that will be exposed to the consequences of these weapons systems? Umbrello attempts to help his arguments by saying that the technological systems must respond reasonably. However, artificial intelligence is designed in relation to rationality, not human reason. Algorithms are rational, more rational than humans can credit. On the other hand, there is little to suggest that they are emotionally or socially intelligent. In this context, one could say that: If everything other than the rational disappears, then human madness has reached its peak. This is where the new technology has the greatest potential to result in ethical collapse. The new technology is becoming so 'rational' that everything is measured, weighed and numbered. If this 'rational logic' is integrated in intelligent robots, then we risk ethical and emotional shipwreck. If this statement is true, or even partly true, then we should be really careful about hoping and believing that so-called rational weapon systems can be ethically and value-wise controlled.

Taking the above reservations into account, Umbrello's book is very important in that it offers insights into technology, ethics and weapon systems. In particular, this applies to weapon systems that are autonomous, semi-autonomous and respond to incoming signals. However, we are of the opinion that no matter how much value design and values are integrated into autonomous weapon systems, the technology will always affect the values so that they are changed when the technology is applied. The opposite model that the values will influence the technology will apply in some cases but not where the consequences of not using such a technology exceed the dangers of using it.

The point of examining and discussing Umbrello's book in this context is to show that there are systemic connections that must be constantly examined so that we do not create hubris solutions that make more problems than they solve.

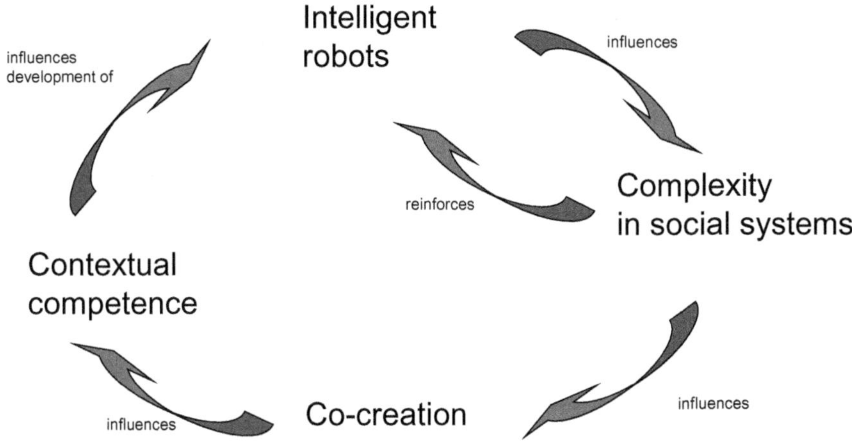

Figure 6.4 Intelligent robots: A loop diagram showing the various relationships.

Intelligent robots, regardless of which field they are used in, increase the complexity of social systems. Consequently, the solution to any problems that may arise is not the punctuation of historical events with the aim of creating simple and clear solutions.[24]

Based on the above description, analysis, case letters, theoretical reflections, the practical utility and the systemic connections, we have developed a loop diagram to show the relationships between intelligent robots, contextual competence, co-creation and complexity Figure 6.4).

Sub-conclusion related to waves of black swan events

When the changes happen faster than we can manage to adapt to them, then perhaps adaptation is not the right strategy! It is possible that a better strategy is to prepare change readiness. In plain language, this means that people, organizations and society need to create readiness for change in order to deal with any situations that can arise. It is the emphasis on the word **can**, which is essential here, not what will come. However, we know with certainty that in the innovation economy new technological innovations will be developed, as well as institutional innovations and other types of innovations as we head towards the Fourth Industrial Revolution. In this context, creating readiness for change means being competent at making changes, managing resistance to changes and coping with the ambivalence that will be a consequence of the changes.

Understanding ambivalence can prove to be just as important as preparing readiness for change when tackling complexity. Social systems are characterized by ambivalence, doubt, belief, emotions, reason and uncertainty, which are all elements that make humans human, and which distinguish us from the rational machines driven by AI technology.

In such social systems, to 'count, weigh, and measure' everything is probably not the best strategy to deal with future changes. When the changes are complex, the strategies one adopts need to have sufficient variety in order to be able to cope with these changes. This is where the Law of Requisite Variety can come into play. Social systems are dynamic and complex. One cannot find out why a bird flock flies in a particular pattern by dissecting the individual birds. It is the flock of birds as a whole that one must study. The point of referring to this analogy is to emphasize that the sum of the parts does not add up to the whole. You must first understand the whole in its context, and then you can examine how the parts and the whole interact.

There is a growing tendency where being 'scientific' means using methods to investigate questions, where the results are presented using tables and figures. The more thorough and well-founded the figures are, the more 'scientific' the research is perceived to be. I call attention to this here, because in many contexts, AI applications can examine a sequence of events that can be quantified more efficiently than humans. However, quantification does not make research results more scientific than qualitative, ambivalent and indeterminate answers. In this connection, we can refer to Karl E. Weick's strategy for studying organizations, which he calls 'organizational theorizing' (1979: 234). Where organizational theorizing becomes central to understanding changes and future workplaces, art will also be important, because: 'art is the attempt to wrest coherence and meaning out of more reality than we ordinarily deal with'.[25] Art, as organizational theorizing, can thus be a method for making complexity meaningful and understandable. An important point regarding this insight in this context is to emphasize that problems and phenomena need to be understood in a larger context than just the bits and parts, numbers and volume in order to find the answers to questions that deal with complex relationships.

Notes

1 Before Australia was discovered by Europeans, it was believed that all swans were white. However, there were black swans in Australia (Taleb, 2010: xi).
2 https://cascadeinstitute.org/wp-content/uploads/2022/04/What-is-a-global-polycrisis-v2.pdf
3 https://futureearth.org/wp-content/uploads/2021/12/GlobalRisksPerceptionsReport2021.pdf
4 North, 1990, 1993.
5 This insight is based on a link between Asplund's Motivation Principle and North's Action Theory. We have formulated Asplund's Motivation Principle on the basis of Asplund's research. In brief, this principle can be described by the following statement: *People are motivated by positive social responses*. The following statement may be said to be one of the central ideas in Asplund's research: *When people receive positive social responses, their level of activity increases* (Asplund, 2010: 221–229).
6 Bateson, 1972: 292–293.
7 Merton, 1967
8 Merton, 1967: 43.
9 Elster, 1986, 1989.

10 Hedstrøm & Swedberg, 1998: 1.
11 Bunge, 1967, 1996.
12 Hernes, 1998.
13 Bunge, 1997: 414.
14 Bunge, 1997: 414.
15 Robot Process Automation is associated with, and inspired by, the thinking of Professor Stafford Beer (Beer, 1979, 1981, 1994, 1995).
16 This is synonymous with Ackoff's 'interactive planning' (1981).
17 The concepts are described in case letter 2 above.
18 https://en.wikipedia.org/wiki/Honey_bee
19 Ashby, 1981; Bostrøm, 2016; Bohr & Memarzadeh, 2020; Carou et al., 2022; Crevier, 1993; Esposito, 2022; Johannessen, 2023; Kumar et al., 2021; Kurzweil, 2008, 2013; Mahajan, 2021; Russel & Norvig, 2016
20 https://en.wikipedia.org/wiki/Hubris
21 https://en.wikipedia.org/wiki/Stanislav_Petrov
22 https://en.wikipedia.org/wiki/Open_Letter_on_Artificial_Intelligence
23 Bateson, G. (2004). *Angels Fear: Towards an Epistemology of the Sacred*, Hampton Press, London.
24 By punctuation (Bateson, 1972: 292–293) a distinction is drawn between cause and effect; this is done with a clear motive in mind. A causality is thus created which does not actually exist in the real world, and one is then free to discuss the effects of this cause which has been created through a process of punctuation.

A sequence of a process is selected and then bracketed. In this way, we de-limit what is punctuated from the rest of the process. Figuratively, we may imagine this as a circle that is divided into small pieces; one piece of the circle is then selected and folded out into a straight line. This results in the creation of an artificial beginning and end. This beginning and end of course cannot exist in a circle, but only through the process of punctuation.

25 Weick (1979: 234) refers here to Peter Vaill.

References

Ackoff, R.L. (1981). *Creating the corporate future: Planned or be planned for*, John Wiley & Sons, New York.

Ackoff, R.L. (1999). *Re-creating the corporation*, Oxford University Press, Oxford.

Ackoff, R.L. (2006). *Idealized design*, FT Press, New York.

Adriaenssen, D.J. & Johannessen, J-A. (2016a). Prospect theory as an explanation for resistance to organizational change: Some managerial implications, *Problems and Perspectives in Management*, 14 (2): 84–92.

Ashby, W.R. (1961). *An introduction to cybernetics*, Chapman & Hall LTD, New York.

Ashby, W.R. (1970). Connectance of large dynamic (cybernetic) systems: Critical values for stability, *Nature*, 228 (5273), 784.

Ashby, W.R. (1981). What is an Intelligent Machine? In Conant, R. (ed.), *Mechanisms of intelligence*, Intersystems, Seaside, CA.

Asplund, J. (2010). *Det sociala livets elementära former*, Korpen, Stockholm.

Bateson, G. (1972). *Steps to an ecology of mind*, Intex Books, New York.

Beer, S. (1979). *The heart of enterprise*, John Wiley & Sons, Chichester.

Beer, S. (1981). *Brain of the firm*, John Wiley & Son, New York.

Beer, S. (1994). *Beyond dispute: The invention of team syntegrity*, Wiley, New York.

Beer, S. (1995). *Diagnosing the system for organizations*, John Wiley & Sons, London.

Benyus, J. (2002). *Biomimicry: Innovation inspired by nature*, Avon, New York.

Bohr, A. & Memarzadeh, K. (2020). *Artificial intelligence in healthcare*, Academic Press, New York.

Bostrøm, N. (2016). *Superintelligence*, Oxford University Press, Oxford.
Bunge, M. (1967). *Scientific research, vol. 3, in studies of the foundations methodology and philosophy of science*, Springer Verlag, Berlin.
Bunge, M. (1996). *Finding philosophy in social science*, Yale University Press, New Haven.
Bunge, M. (1997). Mechanism and explanation, *Philosophy of the Social Sciences*, 27: 410– 465.
Carou, D.; Sartal, A. & Darim, J.P. (2022). *Machine learning and artificial intelligence with industrial applications: From big data to small data*, Springer, London.
Crevier, D. (1993). *AI: The tumultuous search for artificial intelligence*, Basic Books, New York, NY.
Davenport, T.H. (2005). *Thinking for a living, how to get better performance and results from knowledge workers*, Harvard Business School Press, Boston.
Davenport, T.H. (2019). *The Ai advantage*, The MIT Press, Cambridge, MA.
Drucker, P.F. (1999). *Knowledge worker productivity: The biggest challenge*, California management Review, 41 (2): 79–94.
Drucker. P.F. (1999a). *Management challenges for the 21st century*, Harper Collins, New York.
Drucker, P.F. (2005). Managing oneself, *Harvard Business Review*, January: 100–109.
Drucker, P.F. (2007). *Innovation and entrepreneurship*, Elsevier, London.
Eco, U. (1992). *How to travel with a salmon, and other essays*, San Diego, Harcourt.
Eco, U. (1994). *Six walks in a fictional wood*, Harvard University Press, Cambridge, MA.
Eco, U. (2000). *Kant and the platypus: Essays on language and cognition*, Harvest Books, New York.
Eco, U. (2002). *On literature*, Harcourt Books, Orlando.
Eco, U. (2003). *Mouse and rat? Translation as negotiation*, Orion Books, London.
Elster, J. (1986). *Rational choice*, New York University Press, New York.
Elster, J. (1989). *Nuts and bolts for the social sciences*, Cambridge University Press, Cambridge.
Esposito, E. (2022). *Artificial communication: How algorithms produce social intelligence*, The Mit Press, Boston.
Harman, , J. (2013). *The Shark's Paintbrush*, Nicholas Brealey Publishing, New York.
Hedstrøm, P. & Swedberg, S.R. (1998). Social mechanisms: An introductory essay, in Hedstrøm, P. & Swedberg, R. (eds.), *Social mechanisms: An analytical approach to social theory*, Cambridge University Press, Cambridge, pp. 20–32
Hernes, G. (1998). Real virtuality, in Hedstrøm, P. & Swedberg, R. (eds.), *social mechanisms: An analytical approach to social theory*, Cambridge University Press, Cambridge, pp. 74–102.
Johannessen, J-A. (2020). *The workplace of the future*, Routledge, London.
Johannessen, J-A. (2020a). *Automation, innovation and economic crises: Survival the fourth industrial revolution*, Routledge, London.
Johannessen, J-A. (2020b). *Artificial intelligence, automation and the future of competence at work*, Routledge, London.
Johannessen, J-A. (2020c). *Knowledge management for leadership and communication: AI, innovation and the digital economy*, Emerald, London.
Johannessen, J-A. (2020d). *Knowledge Management Philosophy: Communication as a strategic asset in knowledge management*, Emerald, London.
Johannessen, J-A. (2021). *Artificial intelligence, automation and ethics in the innovation economy*, Routledge, London.
Johannessen, J-A. (2021a). *China's innovation economy: Artificial Intelligence and the new silk road*, Routledge, London.
Johannessen, J-A. (2022a). *Creativity, innovation and the fourth industrial revolution: The da Vinci strategy*, Routledge, London.

Johannessen, J-A. (2022b). *The new silk road and the innovation economy in China*, Routledge, London.
Johannessen,J-A. (2023). *Intelligent robots consciousness and creativity: The search for hidden knowledge, The cognitive side of knowledge management*, Emerald, London.
Johannessen, J-A. (2023a). *Feudal capitalism in the innovation economy*, Routledge, London.
Johannessen, J-A. (2023b). *The new silk road and the innovation economy in China*, Routledge, London.
Johannessen, J-A. (2023c). *De-Globalization in the innovation economy:* ***Technological innovations trigger conflict between China and the United States***, Routledge, London.
Kahneman, D., & Tversky, A. (1979). An analysis of decision under risk, *Econometrica, Journal of the Econometric Society, 47*(2), 263–292.
Kahneman, D., & Tversky, A. (2000). Prospect theory: An analysis of decision under risk, in Kahneman, D. & Tversky, A. (eds.), *Choices, values and frames*, Cambridge University Press, Cambridge, pp. 17–43.
Kanter, R. M. (1983). *The change masters*, Unwin Hyman, London.
Kumar S., Aggarwal, V. & Gupta, S. (2021). Artificial Intelligence and Nanotechnology: A super convergence, In Bhargava, C. & Kumar Sharma, P. (eds.), *Artificial intelligence: Fundamentals and applications*, Routledge, London, pp. 1–9.
Kurzweil, R. (2008). *The age of spiritual machines: When computers exceed human intelligence*, Penguin, London.
Kurzweil, R. (2013). *How to create a mind: The secret of human thought revealed*, Penguin Books, New York.
Mahajan, P.S. (2021). *Artificial intelligence in healthcare*, MedMantra, London.
Merton, R.K. (1967). *Social theory and social structure*, Free Press, London.
Miller, J.G. (1978). *Living systems*, McGraw-Hill, New York.
North, D.C. (1990). *Institutions, institutional change and economic performance*, Cambridge University Press, Cambridge.
North, D.C. (1993). *Nobel prize lecture*. http://www.nobelprize.org/nobel_prizes/economics /laureates/1993/north-lecture.html#not2, date of reading: 4.5.2021.
Pask, G. (1992). *Interaction of actors: Theory and some applications*, vol. 1 (unpublished manuscript). www.cybsoc.org/PasksIAT.PDF.
Pfeffer, J. & Sutton, R.J. (1999). Knowing what to do is not enough: Turning knowledge into action, *California Management Review*, 42 (1): 83–108.
Pickering, A. (2010). *The cybernetic brain*, The University of Chicago Press, Chicago.
Russel, S. & Norvig, P. (2016). *Artificial intelligence: A modern approach*, Pearson, New York.
Schumacher, E.F. (1993). *Small is beautiful*, Vintage, London.
Searle, J.H. (1995). *The construction of social reality*, Penguin, London.
Seeley, T. (2019). *Lives of Bees: The untold story of the Honey Bee in the wild*, Princeton University Press, Princeton.
Senge, P.M. (1990). *The fifth discipline. The art and practice of the learning organization*, Doubleday/Currency, New York.
Sennett, R. (2013). *The rituals, pleasures and politics of cooperation*, Penguin, London.
Susskind, D. (2020). *A world without work: Technology, automation and how we should respond*, Allen Lane, London.
Taleb, N.N. (2010). *The Black Swan*, Penguin Books, London.
Umbrello, S. (2022). *Designed for death: Controlling killer Robots*, Trivent Publishing, London.
Watzlawick, P. (1988). *Ultra-Solutions: How to fail most successfully*, W.W.Norton & Company, New York.

Watzlawick, P.; Weakland, J.H. & Fisch, R. (2011). *Change*, W.W.Norton & Company, New York.
Weick, K.E. (1979). *The social psychology of organizing*, Wiley, New York.
Wiener, N. (1948). *Cybernetics; or, control and communication in the animal and the machine*, MIT Press, Cambridge, MA.
Zuboff, S. (2019). *The age of surveillance capitalism*, Profile Books, London.

Epilogue

When systemic innovations overwhelm social systems, we can see the challenges facing us but not the factors that drove the innovations in the first place. On the whole, we can see the same process at work when we consider artificial intelligence and intelligent robots. We see the results but not the drivers that were crucial in bringing them about. The lack of transparency that conceals the drivers is what causes their results to be perceived as random. This also fuels the perception of uncertainty. When we can neither see nor understand the drivers, our imaginations run riot with ideas about how the consequences of these changes may affect our futures.

One of the results is an excessive belief in facts and information that can be verified. The point is that if we try to verify what cannot be verified, we create models of an imagined future. The paradox is that these models do not just become models of an imagined future but models for a future. What we face when the changes take this form is a type of pathological thinking. This thinking takes the form that we first uncover what we believe will be a model for our future. Then, precisely this future is created when we apply these models to social systems. Then at a point in the future, we analyse the social system and discover that the model we first used has become a social reality. In other words, models of the future become reality. We therefore create what we believe will exist in the future and then verify this assumption later when it is shown to be correct. The point of this brief philosophical reflection is that the models are not developed evolutionarily from practice, but practice becomes the result of the models. Metaphorically, one can say that the map changes the terrain, precisely because we apply this map to the terrain.

An underlying point here is that the future cannot be predicted when the rate of change is very great. When the future is created by interconnected unforeseeable events, that is, by waves of black swans, then it is difficult to gain an overview of what this future will look like. These events appear as clusters of events, which are interconnected and therefore almost impossible to gain knowledge about using classical linear models and classical causal thinking. If we isolate events that are connected, we will of course get clear answers. These responses can be measured, weighed and counted. They appear to be possible and relatively certain. They are perhaps correct but based on the wrong way of thinking. In other words, we apply cause-effect models to events that are not of this type but circular with built-in

DOI: 10.4324/9781003567325-7

feedback loops; this can result in dramatic errors that can possibly damage the systems being examined.

It is not the case that systemic thinking denies the principle of cause and effect. However, we need to be aware of creating cause and effect by means of punctuation, that is, creating causality by separating a sequence of a process and bracketing it. In other words, using this method of causality is not always appropriate in relation to the problem being examined. This can result in creating a 'false' future by drawing a map (plan) of the terrain (the future), and then using the map to change the terrain. Thus, it is how you draw this map that also determines what the terrain will look like. This explains why systemic thinking should be preferred when planning the future. Systemic thinking attempts to make sense of the complexity of society by viewing it in terms of parts and wholes, and the relationships between the parts, rather than separating the parts, and applying the principle of punctuation and cause and effect. Using systemic thinking enables us to develop and draw maps (plans) to facilitate change.

Why do we think in linear causal terms, when we know that the picture of social causality is much more complex? Perhaps one reason for such erroneous thinking is that it requires much effort to collect information that shows how the systemic causal links work. The information can often be found in our thoughts and perceptions. It is only signals and indications that can be detected in such linked systemic relationships. In other words, the 'information' that is being sought out does not exist in concrete terms, so people often look for something that can be found in a completely different place. Analogously, this may be compared to the anecdote about the 'Policeman and the drunk man':

> A policeman sees a drunk man searching for something under a streetlight and asks what the drunk has lost. He says he lost his keys and they both look under the streetlight together. After a few minutes the policeman asks if he is sure he lost them here, and the drunk replies, no, and that he lost them in the park. The policeman asks why he is searching here, and the drunk replies, 'this is where the light is'.

This is also the case with the information we are looking for. It cannot be found where we are looking for it, because it has not materialized as information.

So what does this brief philosophical reflection have to do with the fact that we are unable to adapt to changes that occur at a rapid pace? It tells us that erroneous thinking, which is here related to logical cause-and-effect thinking, is a 'hidden' factor, because everyone believes it is the only right way to think. It is 'hidden' because it is the obvious way of thinking. Even if we go over the cliff edge using this type of thinking, the high priests of this type of thinking will say that it was inevitable, because this was the way we were headed. Madness has many facets, and using the principle of natural causality to explain how social systems work is one of the facets.

Index

For Product Safety Concerns and Information please contact our EU representative GPSR@taylorandfrancis.com Taylor & Francis Verlag GmbH, Kaufingerstraße 24, 80331 München, Germany

Batch number: 10397794

Printed by Printforce, the Netherlands